Wireless Networks
FOR
DUMMIES®

by Barry Lewis and Peter T. Davis

WILEY

Wiley Publishing, Inc.

Wireless Networks For Dummies®

Published by
Wiley Publishing, Inc.
111 River Street
Hoboken, NJ 07030-5774
www.wiley.com

Copyright © 2004 by Wiley Publishing, Inc., Indianapolis, Indiana

Published by Wiley Publishing, Inc., Indianapolis, Indiana

Published simultaneously in Canada

For general information on our other products and services or to obtain technical support, please contact our Customer Care Department within the U.S. at 800-762-2974, outside the U.S. at 317-572-3993, or fax 317-572-4002.

Wiley also publishes its books in a variety of electronic formats. Some content that appears in print may not be available in electronic books.

Library of Congress Control Number: 2004107896

ISBN: 0-7645-7525-2

Manufactured in the United States of America

10 9 8 7 6 5 4 3 2 1

1O/SS/QZ/QU/IN

WILEY

About the Author

Barry D. Lewis (CISSP, CISM) has been in the information technology sector for 35 years, specializing in information security since 1980. He co-founded Cerberus Information Security Consulting in 1993 and was elevated to President of the firm shortly thereafter. He served as Secretary and then Vice President of the ISC(2) organization, past-President of the Toronto ISSA, and Chairperson of the Eastern Canada ACF2 Users Group. Barry is listed in the *International Who's Who of Entrepreneurs.* Mr. Lewis lectures worldwide on computer security and audit issues and was honored with the Best Speaker Award at the annual Computer Associates Enterprise Wide Security & Audit conference (ESAC), one of North America's largest security conferences. Mr. Lewis has written and co-authored numerous articles and seven books, including *Computer Security For Dummies,* and *Teach Yourself Windows 2000 Server in 21 Days.* He lives near Toronto with his wife Elizabeth and son Derek.

Peter T. Davis (CISA, CMA, CISSP, CWNA, CCNA, CMC, CISM) founded Peter Davis+Associates (a very original name) as a firm specializing in the security, audit, and control of information. A 29-year information systems veteran, Mr. Davis's career includes positions as programmer, systems analyst, security administrator, security planner, information systems auditor, and consultant. Peter also is the past President and founder of the Toronto ISSA chapter, past Recording Secretary of the ISSA's International Board, and past Computer Security Institute Advisory Committee member. Mr. Davis has written or co-written numerous articles and 10 books, including *Computer Security For Dummies* and *Securing and Controlling Cisco Routers.* Peter was also the technical editor for *Hacking For Dummies* and *Norton Internet Security For Dummies.* Peter is listed in the *International Who's Who of Professionals.* In addition, he was only the third Editor in the three-decade history of *EDPACS*, a security, audit and control publication. He finds time to be a part-time lecturer in data communications at Seneca College (`cs.senecac.on.ca`). He lives with his wife Janet, daughter Kelly, two cats, and a dog in Toronto, Ontario.

Dedication

To my wife Elizabeth, who puts up with far more than I have a right to expect.

— Barry

To all my friends and enemies. Hopefully, the first group is bigger than the second.

— Peter

Author's Acknowledgments

We'd like to offer special thanks to Pat O'Brien, who started this rolling. Peter worked as the technical editor for Pat on *Hacking For Dummies.* Thanks for passing on Melody's name.

Thanks to Melody Layne, acquisitions editor, for pitching the book to the editorial committee and getting us a contract. Much appreciated.

Thanks to Becky Huehls, who started us out on this project as editor but wisely got herself re-assigned. Thanks to Kelly Ewing for picking up the ball and running with it after Becky. Unfortunately, Kelly fumbled it, but Colleen Totz was able to struggle with it over the goal line.

Dan DiNicolo, technical editor, is commended for his diligence in reviewing the material. Thanks, Dan.

Peter would like to thank Kevin Beaver, Ken Cutler, Gerry Grindler, Ronnie Holland, Carl Jackson, Ray Kaplan, Kevin Kobelsky, Carrie Liddie, Dexter Mills Jr., and Larry Simon for responding to a request for wireless information. Thanks for answering the call for help. The provided information shows in this book.

Barry would like to thank his co-author Peter. Always a pleasure, sir. He would also like to acknowledge Craig McGuffin and John Tannahill who are always there for him, as friends and business associates , and never fail to lend a helping hand.

Publisher's Acknowledgments

We're proud of this book; please send us your comments through our online registration form located at www.dummies.com/register.

Some of the people who helped bring this book to market include the following:

Acquisitions, Editorial, and Media Development

Project Editor: Colleen Totz, Rebecca Huehls

Acquisitions Editor: Melody Layne

Technical Editor: Dan DiNicolo

Editorial Manager: Carol Sheehan

Media Development Manager: Laura VanWinkle

Media Development Supervisor: Richard Graves

Editorial Assistant: Amanda Foxworth

Cartoons: Rich Tennant, www.the5thwave.com

Production

Project Coordinators: Courtney MacIntyre, Erin Smith

Layout and Graphics: Jonelle Burns, Lauren Goddard, Denny Hager, Joyce Haughey, Stephanie D. Jumper, Michael Kruzil, Lynsey Osborn, Jacque Roth, Heather Ryan, Ron Terry

Proofreaders: John Greenough, Carl W. Pierce, Brian H. Walls

Indexer: TECHBOOKS Production Services

Special Help: Kelly Ewing

Publishing and Editorial for Technology Dummies

Richard Swadley, Vice President and Executive Group Publisher

Andy Cummings, Vice President and Publisher

Mary Bednarek, Executive Acquisitions Director

Mary C. Corder, Editorial Director

Publishing for Consumer Dummies

Diane Graves Steele, Vice President and Publisher

Joyce Pepple, Acquisitions Director

Composition Services

Gerry Fahey, Vice President of Production Services

Debbie Stailey, Director of Composition Services

Contents at a Glance

Table of Contents

Introduction

●●●

*R*ecently, a very knowledgeable speaker at a presentation for a wireless vendor talked about wireless as ubiquitous. We would have to disagree. Wireless is widespread, but it is not everywhere. But it is rapidly becoming ubiquitous.

In about 1990, cell phone users carried around a phone that looked and felt like a World War II walkie-talkie. You didn't casually whip that baby out and start a conversation. At that time, you either had a deep wallet, a big ego, or a compelling need to talk to your mother. Now, depending on where you live in the world, the cell phone provides better quality at a lower cost — and in a much smaller form factor.

In about 1994, Wireless Local Area Network (WLAN) equipment manufacturers sold gear that was comparable to the wireless phone market. The devices were large and very expensive and provided poor bandwidth. You had to have a compelling reason to spend the money on the low bandwidth. But that has changed. In 2004, you can buy a reasonably priced laptop with onboard support for 802.11b and Bluetooth. You can add a fairly inexpensive 802.11a and g PC Card and easily connect wirelessly and, if you have a Centrino-based system or Microsoft Windows XP, somewhat seamlessly. The WLAN market of today is analogous to the Ethernet market of the mid 1980s. In those years, solutions were proprietary, and standards were being approved. Companies were jockeying for position. Now, you would have difficulty finding a desktop or laptop computer that does not come with Ethernet support. When we look back from the future, we will see some parallels between 802.3 and 802.11 development.

Telephone companies have wrestled with the issue of the "last mile" for a while. But forget the last mile; Bluetooth and IrDA provide the last foot. Mice, keyboards, phones, PDAs, and other devices support Bluetooth for wire replacement. When you have a Bluetooth-enabled printer, you no longer need to hook up to the network to print one page or a contact.

Vendors are trying to get along with each other to develop standards so that one day we can walk around with a phone, PDA, or laptop and connect to any network, anywhere, anytime.

We wrote this book for those of you who want to release your company from its bondage. If you want to unfetter your clients so that they can access their e-mail before getting on the red-eye to New York or Toronto, this book is for you. If your desktop looks like spaghetti junction and you want to rid yourself of all those wires, this book is for you. If you want to provide up-to-the-minute stock quotes to the Chair of your company while she sits in the boardroom, this book is for you. If you have a small to medium enterprise (SME) or business (SMB) or a small office/home office (SOHO) and don't want to rewire (or pay someone to rewire) your office, this book is for you. If you have wireless at home and want to learn about features that you can expect for home gear, this book is for you.

About This Book

Mark Twain once wrote, "Writing is easy. All you have to do is cross out the wrong words." So we have done the easy part for you. We crossed out all the wrong words and came up with this book. What are left are the words that will help you plan for, install, acquire, protect, manage, and administer wireless networks from personal area to wide area.

We have started your journey by providing information on

- Differentiating WPAN, WLAN, WMAN, and WWAN
- Planning your wireless network
- Doing a site survey
- Using IrDA for transferring files
- Using Bluetooth for wire replacement
- Securing IrDA and Bluetooth
- Acquiring the right equipment
- Installing and configuring an access point
- Acquiring and installing client hardware
- Installing and configuring client software
- Building a network to allow roaming
- Connecting while on the go
- Securing your WLAN
- Understanding WEP, WPA, and RSN
- Understanding EAP

　　　✔ Setting up a VPN using PPTP

　　　✔ Troubleshooting your network

　　　✔ Evaluating and fixing network performance

　　　✔ Using administrator software and utilities

How to Use This Book

You don't have to start reading this book at the beginning — each chapter stands on its own, as does each and every part. In fact, if you are new to wireless networking but not new to networking, you may want to start with the radio frequency (RF) primer in Appendix C. It's an exploration of radio frequency. We do suggest, however, that you consider starting at the beginning and reading to the end.

We encourage you to use the white space and the margins of this book. Mark it up and make it your own. Look up the links that we provideto find a wealth of information. You'll want to write in the book because it will become an irreplaceable reference when you cross out words and add others yourself.

How This Book Is Organized

We grouped the chapters of this book into manageable chunks, called *parts*. Each part has a theme. For instance, the first part is like a handshake. We help you get started through planning and acquiring hardware and software. In between, we have some management parts. And the last part is an exchange of data.

Part 1: Planning and Acquiring Your Network

Part I is the foundation required for the world of wireless networks. Chapter 1 breaks down the various types of wireless networks and provides examples of each type. If you're trying to sell wireless within your organization, you will see some benefits of wireless you might want to use. Chapter 2 introduces the necessary planning for a successful implementation. You will see how to do a site survey. Chapter 3 sorts out more terminology and explains the differences between an ad hoc and an infrastructure network. You get the scoop on differentiating between BSS, IBSS, and ESS.

Part II: Implementing Your Wireless Network

Part II starts at 10 meters and moves to 100 meters and then beyond. Chapter 4 gives you an overview of IrDA and Bluetooth and security measures for both. You will understand how to use these technologies to replace wires in your office or on your body. Chapters 5 and 6 move on to wireless local area networking. In Chapter 5, you see how to set up an access point. Chapter 6 talks about connecting Windows 2000, Windows XP, tablet PC, Linux, and Mac OS clients as well as Centrino-based systems. Chapter 7 moves you a little beyond your office through the use of bridges and switches. When you want to roam about your offices, you need to set your system up correctly. You can read how to do this in Chapter 7. Chapter 8 deals with accessing wireless wide area networks. You will find a discussion about hot spots. But more important, you see how to get your e-mail wirelessly while on the road.

Part III: Using Your Network Securely

In Chapter 9, we introduce you to the additional risks of wireless networks — *additional* because you have all the risks of a wired network *plus* those specific to wireless. For instance, signal jamming is not a real risk in wired networks, but it is in wireless networks.

It wouldn't be fair to just enumerate risks without providing some help. Chapter 10 provides a security architecture for your wireless network. You probably have a healthy paranoia when it comes to wireless and its security. Hopefully, Chapter 11 can allay some of your fears — and add some new ones. Chapter 12 helps you set up a secure channel because, quite frankly, who cares if WEP can be broken when you protect your data at a higher level?

Part IV: Keeping Your Network on the Air — Administration and Troubleshooting

Availability is an integral part of a network. So is administration. After we develop a network that is reliable, available, and secure, we must administer it. This includes documenting components. Chapter 13 highlights known wireless problems and practiced solutions. For instance, when you have a near/far condition, move one of the workstations closer. You'll find pithy advice like this in Chapter 13. Chapter 14 discusses bridges and bridging technology. When you have a network, you have people complaining about connectivity and performance. Chapter 15 provides some commonsense solutions to typical wireless problems. Finally, Chapter 16 provides guidance for trying to wrap your hands (and heads) around your network. Whether you have an authorized wireless network or not, you won't want to miss this chapter.

Part V: The Part of Tens

The Part of Tens provides top-ten that lists authors think are interesting. Our Part of Tens is no different. We provide a look at ten indispensable tools for network administrators and tinkerers alike (Chapter 17); ten ways to secure your WLAN (Chapter 18); and ten ways to use wireless in your business (Chapter 19).

Part VI: Appendixes

We also provide some valuable reference material in the Appendixes. Appendix A lists trade associations and user groups for wireless. You can find out whether there is a WUG (wireless user group) in your neighborhood. If you don't have one, start one and send us the information. Appendix B provides information on the 802 standards that you need to know for the wireless market. Appendix C, as mentioned earlier, provides the minimum necessary information that you need to set up a wireless network. If you want more information, refer to Appendixes A and B.

Icons Used in This Book

Throughout the book, we used icons to highlight specific material. If we were there with you in person, we would change the volume or cadence of our speech to get your attention. Because we're not, we use these icons to accomplish the same thing. You will find the following icons used in this book:

This is your clue to get out your pocket protector and calculator. We might as well put a really big technical stuff icon on Appendix C. Appendix C is not for technophobes.

When you see this icon, you'll find something that will save you time and money. Now, we wrote the entire book to do this, but these are the areas we want you to pay particular attention to.

Have you every read the instructions for your microwave oven? (We just can't get away from these frequency illustrations; perhaps we need spectrum analysis.) Some of these instructions say to not put animals in the microwave oven and the turn it on. Or, have you ever been to California and stayed in a hotel where the bathtub had a warning label that said "Warning: fixture slippery when wet"? Well, these are examples of where we would use our warning icon. For example, don't plug your AC device into the DC outlet (without an inverter), or don't put the antenna too close to your head. Do you remember when people were warned about cell phone antennas? How the heck do you use a phone without putting it close to your head? Hands-free, that's the answer. Enter Bluetooth, speakerphone, or an ear bud.

You may find some Remember icons, should we remember to put them in. We might say, "Remember to back up the registry before changing a key." There, we remembered to say it.

Where to Go from Here

Learning is a path, not a destination. This is especially true of wireless network learning because of the rapidly evolving wireless marketplace. When you embark upon wireless networking, you embark on a process of life-long learning. We hope you like change; otherwise, you may want to forget about wireless.

If you are just starting out, start at Chapter 1. If you already have equipment and want to get at it, skip to Chapter 3. If you want to secure your network, start with Part III. It doesn't matter where you start as long as you eventually read the whole book. There is something for everyone in every chapter.

We will try to keep you up-to-date with changes applicable to this book. You can check www.dummies.com/extras for them.

A journey of a thousand miles starts with a single step. So on to Chapter 1!

Part I
Planning and Acquiring Your Network

The 5th Wave   By Rich Tennant

"This part of the test tells us whether you're personally suited to the job of network administrator."

In this part . . .

The foundation provided in this part is required for entering into the world of wireless networks. You discover the elements needed to become an educated user of wireless networks, and you see how planning your use of this technology is a critical component. During your planning, you see how site surveys are essential to your implementation. Finally, you see how to choose the equipment you need. We also introduce the various wireless modes and frequencies to help you match the equipment you purchase to your plan.

Chapter 1

Removing the Tethers: Entering the Wireless World

In This Chapter

▶ Understanding the risks and the rewards of going wireless

▶ Sorting out acronyms and types of wireless networks

▶ Planning and installing your wireless network

▶ Administering and troubleshooting

*N*ow is an exciting time for network administrators and users everywhere as we cast off the shackles of our wired world and move into the new frontier of wireless networking. This book shows you the steps to take to accomplish this as seamlessly and reliably as possible while protecting your corporate assets from unauthorized access.

In many office environments, the desk and the workstation it supports is a fixed entity. Every day you come to the office, sit at your desk, and power on your computer, ready to start the day. There is little other choice because your workstation needs a cable from it to the network in order for your applications and Internet support to work. You've been wired for years.

Perhaps, however, you've seen how wireless networking offers your business an opportunity to move beyond the expense of wires and cables into the less expensive world of air and radio waves. You've seen this already with the introduction of cell phones. Can you imagine waiting to reach someone until they returned to the office or home and their old-fashioned POTS?

What's POTS, you ask? *Plain Old Telephone Service.* Land lines. Cables. Restrictions. In the world today, if you are a teenager, you've never known that it was impossible to reach someone unless he was by his home phone, have you? You've always known about cell phones and probably have one or two yourself. Now you can move your computer into the same wireless world and free yourself from the same restrictions of needing to access a fixed, physical link. You can sit on the front porch and enjoy the sunshine with your laptop wirelessly connected to your local area network and the Internet. This wasn't possible only a few years ago.

Understanding the Risks and Rewards of Going Wireless

Going wireless has wonderful benefits, but wireless freedom comes with its share of problems as well. You need to be aware of these concerns as you move into this fascinating new world. As Oprah Winfrey once said, "I believe that one of life's greatest risks is never daring to risk." What we need to do is take calculated risks, with forethought and intelligent analysis.

What you risk

What types of risks do you face with wireless networking? Are they more or less than the risks you already face with cable-based networks? Chapter 9 provides insight on these risks, with later chapters offering solutions. In a nutshell, the risks in wireless networks are different than in cable-based networks in that physical security, like with door locks and surveillance cameras, helps protect us from the weaknesses in our local area networks. Attackers need to physically connect to the network in order to attack it. Now, I hear many of you saying that this isn't the case, and that dial-up or other remote access points offer vulnerabilities that can be abused. You are right, but this is only one aspect of gaining access to a network; and, if you disallow modems or external access, you are left with physical access as the sole entry point. This is not the case with wireless.

Many of you may recall using television antennae years ago to receive your TV stations. Some of you may still use antennae, especially in the country. This is a wireless model. Pick up an antenna, plug it into your TV, and off you go — free TV! Now you can do that with a computer. Add a wireless modem, an antenna (which may be part of the modem or integrated into the computer), and go looking for wireless signals to connect to and use. Free wireless! Okay, not really. Yes, you can do as we described; however, in many places, using someone else's network is illegal — and doing so will land you in the hoosegow.

In a wireless network, you broadcast your network to the world. "Hello? Here I am. Come and get me." Wireless networks beg to be used (or abused). And this is where they differ from land-line-based networks. Walk on by and see whether you can see a signal and use it. We point out in later chapters how people have been arrested for accessing pornography using the neighborhood wireless access points that were left unsecured and accessible to anyone. You really do not want the police knocking on your door one day asking about illegal network traffic, do you? Well, that is one of the risks of wireless.

Illegal use of your network is your biggest nightmare. This may include the scenario just mentioned or someone hacking into your network to steal commercial secrets. Neither is a good thing. Most of this risk comes from poor design and inadequate use of the security components available in a wireless network. Part III shows you how to properly secure your wireless network from intruders.

The benefits you gain

The benefits of a wireless network can be almost immeasurable. As we mention earlier, can you imagine a world without cell phones now? You almost automatically assume you can reach people any time you need to by calling their cell phone. Now imagine not being tied to your desk to accomplish your work. With the latest tablet PCs, you can roam around the office, from meeting to meeting, tablet PC in your hands, always connected, always available. In addition, you can sit in the cafeteria and grab a coffee and donut while still working on that big proposal. Or enjoy a few days of sun you would ordinarily miss, all while you work diligently away at your job.

Or imagine the usefulness of a wireless connection at the airport while you wait (and wait) for that never-on-time flight. At least now you can do some work, browse the Internet, or connect with other passengers to whittle away the interminable time waiting for flights.

The next few years will bring a revolution in networking, both personal and job-related, as wireless networking becomes *de rigueur*.

Applications of Wireless Networks

So where will you use this fancy wireless networking? Applications abound. We discuss using it in airports and in the office — major uses for any businessperson who travels. The amount of additional work that can be accomplished is immeasurable, hopefully resulting in added responsibilities and increased compensation for you. At the very least, perhaps it offers you more time with your family. What's that, you ask? How's that? Well, consider the work you need to get done each day and the deadlines you have to meet. You now have the time on the train coming home, the time spent traveling, and all that time you used to spend frustrated while waiting for a flight to get your work done so you can be with your family — instead of your work — when you finally arrive home.

The use of wireless networks doesn't end there though. If Bill Gates gets his way, we will see the wireless world in our fridges, stoves, coffee pots, and house alarms. Wait, some of those are already present.

Here are some other uses you may see:

- **Home or small office security:** Wireless cameras can be connected to your Web site, enabling you to visually check in with the office when you are away. We hope that you do this to view the premises after hours, making sure that it remains locked down and isn't broken into. Of course, there is nothing stopping you, local laws notwithstanding, from checking in on your staff while you are away to make sure they aren't partying.

- **Medicine:** Imagine a place where doctors can carry a small tablet PC around and access your records instantly from any location. While there, they might send a prescription directly to the pharmacy, bypassing the need for you to take that scrawl they call *handwriting* to your pharmacy and wait for it to be filled. The nurses could record your vital signs into a wireless device, providing instant access for your physician. Perhaps the doctor will send results to the lab, ask other doctors for advice, and generally serve you better by being fully connected.

- **Live data updates:** Data can be updated live without waiting for staff to return to the office to file paper-based reports on inspections or sales. These staff can just carry their wireless access device and input data, sending it immediately to the main computers for processing, thus speeding up a sales cycle or the collection of data that the business database requires each month.

- **Business applications:** On the business application side, companies like Microsoft, Peoplesoft, and SAP are building wireless into their products, all the better to serve the user and enable faster, more effective data use. Some real estate offices are using wireless to give brokers access to the property-listing database, which can have a dramatic effect on brokers' ability to do their job. For example, suppose a client makes an appointment to look at certain house types, such as bungalows, and the broker meets her with a list of properties fitting that description. Now, suppose that after looking at what's available, the client realizes that she actually wants a two-story home. The broker can immediately dial in a different search rather than rescheduling with the client. Presto, instant sales.

The use of wireless networks will skyrocket in the next few years, we predict, with more and more vendors applying the concept of liberating us from the desk or from manual methods of applying new data to our systems.

Sorting Out the Nets: Do I Need a WPAN, WLAN, or WMAN?

Acronyms are the bane of all professionals. Whether it's a nurse or doctor asking for an MRI or an accountant discussing ROI, you need to know the lingo if you want to be part of the talk. It's the same with wireless networks.

There are a number of different types of networks whose classifications are based primarily on the distances they reach. You see in Table 1-1 how they relate to the wired world and to each other.

Table 1-1	Different Types of Networks	
Network Type	*Wired*	*Wireless*
LAN	IEEE 802.3 (Ethernet)	IEEE 802.11X
PAN	IEEE 1394 USB	IEEE 802.15.1 *Blue Tooth*
	Firewire	IEEE 802.15.3
		IEEE 802.15.4
MAN	Broadband (DSL, cable)	IEEE 802.16

The IEEE (Institute of Electrical and Electronics Engineers) provides standards for everyone to follow. These include standards for wired and wireless networking. The numbers are assigned by the IEEE and quickly become well known to industry users. The 802 series dictates how each format must work. You can obtain lots of interesting information about these standards and their use from various Web sites. One of these is www.dailywireless.org/index.php. This site includes regular updates on what's happening in the wireless world.

Impress your friends by mentioning the 802.11 standard that your wireless network uses. It will be either a, b, or g. 802.11b is currently the most popular, but 802.11g is catching on fast.

Let's get personal: WPAN

The *Wireless Personal Area Network* (*WPAN*) consists of close-range wireless activities such as Bluetooth and FireWire. Wireless in this range is based on the IEEE 802.15 standard. Transmission in this network consists of a low range of around 30 feet, or 10 meters. It's right up there in your personal space, sort of like being in a crowd and getting jostled all the time. It uses low power consumption and is an ad hoc network. If you are in range and another device is present, you can reach out and touch it.

This spectrum is designed for interpersonal connections, such as connecting one PDA to another, or connecting a wireless keyboard, mouse, or printer to your computer. It is useful and helps free you from all the cables typically needed to perform these tasks. Data transfer occurs at around 1 Mbps in the Bluetooth protocol.

Many of you already indulge in the Wireless Personal Area Network world with your infrared-equipped PDA that you use to beam information to other PDA users. Others of you wander around airports with a Bluetooth-equipped headset on your cell phone. That we like. It's got to beat having that darn wire hanging around your neck although we are not sure you realize how many people initially think you are talking to them as you chat away.

Other neat uses of this spectrum include connecting your PDA to your workstation or laptop to synchronize data or adding a Bluetooth-enabled modem like those available from Zoom Telephonics. Why a Bluetooth modem? Well, if you travel, you can connect in your hotel with a dial-up line. Although many hotels are moving to wireless connectivity, many have not gotten there yet. So using a Bluetooth modem provides you a degree of that wireless connectivity as you roam around your hotel room or even step out onto the balcony — all while remaining connected.

Another use of this spectrum regards connecting with other laptop users to share files easily and quickly without the need for network cards or cables. Although you can also do this in the WLAN technology, using infrared allows you to quickly share files with another user, with little fuss and bother.

One quick note of clarification: Personal Area Networks (PAN) actually refer to using a near field electric field to send data across various devices using the body as a medium. The term was really meant to be used as it is in Wireless Personal Area Networks. Nonetheless, it is an accepted term now and is used interchangeably. If you want to read an article describing this original use, visit `www.wirelessdevnet.com/channels/bluetooth/features/pans.html`, a wireless developer's Web site.

The holy grail of wireless networking: WLAN

WLAN is the holy grail of wireless networking for most of the business world. Using the IEEE 802.11 standard, it is the main topic for most of this book. *Wireless Local Area Networking (WLAN)* gets you connected to the office with your laptop or tablet PC, allowing you to roam around at work while remaining connected. It won't be long before you're able to tell the boss you are working while standing around the water cooler and chatting with your friends. Look boss, I'm downloading that latest spreadsheet and discussing it with Harry while slaking my thirst.

This is where *Wi-Fi* — a term sometimes used interchangeably for the IEEE 802.11 standard — originates. This wireless connectivity expands beyond the area of our desks and moves us to further distances. Distances of up to 500 feet are possible with no interference, and even farther distances can be easily achieved using repeaters and additional access points. We guide you through understanding the protocols, the risks, implementing security, and more throughout this book.

You are probably most interested in WLAN, and so we focus on it in this book. A Wireless Local Area Network is analogous to the wired local area network you use each day at the office or even in your home. It is merely a connectivity of devices, allowing sharing of resources. Where WLANs are useful, though, is in freeing up your installation from cable worries and knocking holes in walls and floors to run those connecting cables.

This can be especially beneficial for small and medium businesses (SMBs) because you may not own the rights to your building and so may need to contract with the building maintenance people to get things done. This can be time-consuming and expensive. A WLAN, properly configured and secured, offers unprecedented access with none of those concerns. You can use a WLAN to extend your network beyond the walls and floors of your building, perhaps into an adjoining conference room or lunchroom, providing staff with access while they grab a quick bite to eat.

There are numerous considerations, of course, and security is paramount. Properly configured, however, WLAN can set you free and let you explore areas of connectivity you may never have considered.

Where the rubber hits the road: WMAN

The *Wireless Metropolitan Area Networks,* or *WMAN,* are also sometimes referred to as *Wi-Max* and *WirelessMan.* This is where the rubber hits the road almost literally as the distances reached are far greater than with the prior standards. Based on the IEEE 802.16 standard, the WMAN provides for large distances and high speeds of access. This standard focuses on the efficient use of bandwidth in the 10–66 GHz range, although an amendment to it, 802.16a, allows for access in the 2–11 GHz range.

WMAN offers wireless access to buildings through the use of external antennae accessing central base stations. WLAN is great for one building or a few floors, but if you are a large organization with geographically separate buildings, you might need the extended distance that WMAN offers. An advantage of this protocol is the allowance for quality of service (QoS) to further enhance its use in business. This allows a reseller to guarantee a certain level of service. Of course, you might just buy space on such a network from any of the vendors putting WMANs in place. For example, in Portland, Oregon, VeriLAN Inc. (www.VeriLAN.com) began offering such a service in 2004.

One newer standard includes 802.20, providing connectivity speeds of up to 1 Mbps for vehicular traffic traveling at speeds of up to 250 kilometers (155 miles) per hour. This is cool: Not only can you illegally speed down the highway, but you can interface with a wireless network while doing it. Of course, I suspect the police are more interested in this particular band.

Using Wireless Networks

Using a wireless network takes a number of components and some fairly critical thinking up front before allowing anyone to connect. We discuss these components in the next few sections.

Accessing networks

You need tools to access a wireless network. You also need to be aware of the distances and transmission speeds you want to use in order to choose the correct technology. To quickly summarize, there are a number of competing wireless standards to consider. Table 1-2 covers the more popular ones.

Table 1-2	Popular Wireless Standards
Standard	*What It Means*
802.11a	54 Mbps speed in the 5 GHz band.
802.11b	11 Mbps transmission in the 2.4 GHz band.
802.11g	54 Mbps; remains backward-compatible with 802.11b.
802.15	Personal Area Network standard. Bluetooth is the typical name.

Many other standards exist. You can find a list of them all in Appendix B, but the ones covered in Table 1-2 are the current popular ones. To use them, your network card must support that standard, along with, of course, your wireless access point. After you add a wireless network card to your machine or PDA, you are off to the races and can enjoy mobility while remaining connected to your network.

Depending upon which wireless standard you use, your roaming can include your office floor or perhaps even the entire building, including part of your building's parking lot. We talk more about the pitfalls of that in later chapters.

Extending the network

Now that you are connected, it may occur to you that you need that degree of access in other locations. It's catching, this freedom. To accomplish this, you need to extend the network or extend your ability to reach the current network.

One easy method to extend your range is to improve the antenna you use. Typically, your network card uses a small antenna either right in the card or, with a Centrino or AirPort chipset, somewhere within the computer. These work great for the typical distances involved in your wireless network but fall short of anything substantial. Adding a high-gain antenna to your computer or PDA significantly improves your ability to reach the network from far greater distances.

The alternate method is to extend your network using repeaters and additional access points. This is where planning is effective to ensure that you understand your needs before you start and then implement the technologies necessary to manage those needs.

Connecting buildings

Perhaps you want to ensure that all the buildings your business uses are connected on one seamless network using wireless frequencies. There are a couple of methods you can use for this, including some really advanced methods if your buildings are a great distance apart.

Connecting networks based in separate buildings leads to major benefits, such as users accessing necessary data residing in central resources crossing large open spaces (office complexes and university campuses, for example).

Two common methods for bridging this gap include point-to-point and multipoint LAN bridges. Using these techniques, your wireless network can expand from one room, to one building, to multiple buildings, to across a city. This is a complex implementation, however, and is not likely something the target SMB this book is designed for will use. You can see in Figure 1-1 how each bridge works.

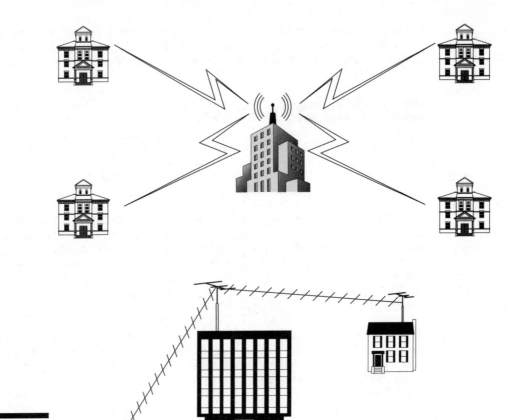

Figure 1-1:
Point-
to-point
(top) and
multipoint
(bottom)
network
bridges.

Going mobile

As the song indicates, we're on the road again. We travel a lot, and getting access to our e-mail and office is essential. Most of the time, we gain access by using a local phone call to our network service provider and then using that access to cross over to our office and get connected.

For security purposes, Barry uses an encrypted tunnel to access his home office. After he's connected, he can easily access all his machines and obtain whatever files he needs. This way, everything he does while connected is protected from prying eyes.

One thing that is changing is the reduced need to use a land-line-based service provider like AT&T Global Services. As hotels, airports, and coffee shops add wireless access hotspots, it gets easier and easier to connect. As the familiar advertising phrase goes, "Can you hear me now?" The answer, increasingly, is *yes*.

 There are numerous utilities, like Boingo, that offer wireless access around the world in coffee shops, restaurants, and hotels. Getting connected while mobile is often simple. Boingo offers free software that finds wireless network hotspots and makes the connection for you. You pay a fee to Boingo to access any of their *hotspots* around the world.

Getting mail on the road

One obvious need for mobile travelers is to retrieve their e-mail. Where would we be without it? This too is becoming simple. Years ago, Barry's business associate would carry alligator clips and a long telephone cable with stripped wires at one end. This was so he could connect to telephone systems in foreign countries where data access was limited or not even considered. Although this has changed and almost every hotel telephone system allows and accommodates telephone modem access today, wireless is also beginning to intrude and become the norm. That certainly is easier than ripping wires out of a wall to get connected!

After you have access to the Internet using a wireless access point, obtaining your e-mail is trivial. It is still ideal to use a Virtual Private Network (VPN) so that your e-mail and passwords are not traversing the network in clear text but traveling through an encrypted tunnel instead. This is even truer in the wireless world because those networks are vulnerable to anyone on the same network sniffing the traffic and seeing what you are up to at any given moment.

One other method involves using a PDA, such as the new Treo 600 or a Blackberry, to obtain e-mail. As data travels over the cell phone's GPRS network, it is slightly less vulnerable than a Wi-Fi connection. It's also very convenient: You merely turn on your device, connect to the local cell provider, and presto! You have mail!

Turning a Notion into a Network

Okay, so you are captured by the possibilities and want your own wireless network. As a small business owner, you cannot afford to hire a third party to install and maintain this network, so you need to understand how to accomplish such a thing by yourself.

It is one thing to desire something and quite another to obtain it in a useful, secure manner. You must take certain steps to protect your business and your wireless investment; *planning,* that awful bugaboo for many of you, is absolutely necessary.

Planning your wireless network

In Chapter 2, you find out all about creating a plan for your new wireless network. We cannot stress this enough: *Do not skip that chapter.* Implementing a wireless solution may be as simple as adding an access point onto your network and letting your staff connect.

But there are pitfalls even with this simple approach. Where will you place the access point? Far too many organizations place them inside the network, which is the absolute wrong place for a wireless connection to be. Your network needs to be protected from any potential wireless attacks; therefore, the access needs to be on the outside of your firewall, forcing users to authenticate their identities to gain access to the internal network.

Where will the wireless access be needed? It makes little sense to place it in the main office if attenuation from the building and its occupants results in the signal not reaching the intended audience. Finally, you need to configure the necessary degree of security to ensure your access is used only by authorized users.

Installing your wireless network

Depending on the size of your wireless network, installation may be as simple as placing an access point on a table or wall and plugging it into a power supply. However, you may also install a more complex system, using repeaters, bridges, and external antennae. These need careful placement and subsequent installation to ensure they meet all your needs and allow for flawless connectivity.

After you plan the installation, it is necessary to begin installing the components. When you do so, you want to follow some structure in order to make the implementation smooth. First, review your plan and ensure that it is complete. Next, unpack the equipment you plan to install and ensure that all the parts are there and that nothing looks broken. Now, connect all the pieces. For an access point, this usually means adding the external antennae that came with the device. However, perhaps you are installing high-gain external antennae and they are to be located on a rooftop. Which comes first, the chicken or the egg? Install the antennae and cabling and then connect it to an access point.

Continue installing access points or repeaters as per your plan until you finish. Make sure that you install wireless network cards in a few worksta-tions or laptops so that you can test accessibility after you configure and secure the network. After all the hardware is in place, you need to configure the network.

Configuring a wireless network

After installing all the access points, you must configure the network. Con-figuring the network sets up the software and all its components so that a wireless signal is transmitted clearly and is accessible to your network cards.

Configuration includes a number of activities. These include setting up the basic parameters that allow your access point and network cards to commu-nicate, thus starting your progress into the wireless world. Other items include those shown in Table 1-3.

Table 1-3	Configuring Your Wireless Network
Parameter	*Description*
Set your IP address.	You need to set the IP address in your network card so it can recognize the access point.
Test connection with the ping command.	Use this command to ensure that you can reach the access point.
Enter the Administration menu.	To set the device parameters, you need the main menu of the device. You enter the vendor-supplied default account and password to accomplish this action.
Set the options.	You need to set the time, disable remote access, deter-mine whether you need DHCP, and ensure that the IP addressing is appropriate for your needs.
Update to the latest firmware.	This is important. Make sure that you follow directions and visit the vendor Web site to get the latest firmware. This ensures that your device is up-to-date and all vendor patches are implemented.

Configuration allows your devices to connect to each other and, if appropriate, with your Local Area Network. After this is established, you need to ensure that your connections are secure.

Staying secure in the wireless world

Securing your network is the most important part of your wireless journey. Don't skip past it in your excitement at being connected to a wireless network. There are many risks to your network, your users, and your data in this new wild, wild west. Risks involve strange names such as war driving and war flying. You didn't know you were getting into a special arcane world of warfare did you?

War driving and *war flying* are exercises in which someone drives or even flies around, equipped with special software, a laptop with a wireless network card, and an external antenna. Using this equipment, they will find your wireless network and probe it to see whether you are using security. You offer an open door when you've skipped those steps and no security is in place.

Other risks include identity theft and data loss. Using that unsecured wireless access point, intruders steal information like credit card numbers, addresses, and even pass codes if you keep these on a computer somewhere on your network. They may even take the special fried chicken recipe you are working on to combat KFC's if you don't secure it well.

Fortunately, there are things you can do to prevent security breaches, or at least to make it exceedingly difficult to break into your network. It starts with turning on encryption and using techniques like Media Access Control (MAC) filtering and even more advanced authentication techniques like Extended Authentication Protocols (EAP) to ensure that only authorized users connect to your network. Finally, you can really improve access security by using techniques called Virtual Private Networking (VPN). We guide you through all these using step-by-step procedures and detailed discussions in later chapters.

Administering and maintaining a wireless network

After your network is set up securely, you'll want to use it all the time. Why not? That is one reason for implementing a wireless network, to set yourself free to wander with your machine, remaining connected as you walk to the conference room or sit in the park.

All this comes at a price, however, because nothing is permanent, and it all requires some degree of administration and support. Depending on the size of your client base, using a security technique such as MAC filtering can be very time-consuming. You need to keep lists of all the MAC addresses used and the corresponding individual network cards in order to track their use and change them when users' network cards fail or laptops change hands and no longer require access.

In addition, troubleshooting any sort of network requires constant surveillance and analysis. In the wireless world, there are issues such as changing *Fresnel zones,* where objects block your signal. Other issues needing constant maintenance might include free space loss, in which changing weather might cut off a fringe signal. And, of course, you need to be aware of typical and abnormal traffic loads. Users suddenly downloading copious quantities of files (they wouldn't be downloading music, would they?) can cause the network to slow to a crawl. Someone needs to monitor and ensure that steps are taken to limit such slowdowns to keep everyone happy.

Throughout this book, we provide a number of tools and several techniques for managing your wireless network after it is up and running. You must keep those happy faces that all your users received when they first signed on to the wireless world and found that freedom.

Convergence of Wireless Technologies — What Will the Future Hold?

Where will we all be in the years to come? No one really knows. We can take educated guesses, though. We are already seeing a huge increase in the use of wireless technologies. Where just a few years ago we would check into the hotel, locate the telephone, and plug in our modem, we now look for a wireless connection first. Barry uses his Treo 600 to send and retrieve e-mail, call home, and search the Web.

This is one area where wireless convergence will skyrocket in the future. We anticipate that all major hotels will be completely wireless in the next three to five years. According to a survey of Internet trends by Ipsos-Insight, it seems that wireless Internet usage grew 145 percent in 2003 with 79 million unique visitors. The study claims that roughly 40 percent of people with land-line Internet access have tried wireless networks. We can expect to see even these figures surpassed in the coming years.

At the airport, your connection will be announced over the wireless network, informing you of delays or arrivals as they occur. No longer will you hang around wondering what is going on when your plane is late, hoping some harried airline staffers will stop to actually consider their customers for a change. (I know — after all the travel Barry does, he still gets upset at the often-cavalier attitude he encounters from airlines.)

Wireless connectivity will continue to grow and become ever more intrusive in our lives. Look for wireless security systems for home and business to grow, coupled with instant messaging and Web page photos to provide greater security and faster notice of break-ins. This can ease the burden of getting up at 2 a.m. to respond to an alarm at the office. Perhaps in the next few years, you'll merely log on and check out the remote cameras to verify whether a break-in occurred before getting dressed and venturing forth. A friend of Barry's installed a Web-based camera at his cottage recently. He can now log on to the Internet, access his Web site, and check for snowfall or intrusions online. That's awesome; his cottage is a two-hour drive away.

Other interesting thoughts include an expansion of the wireless spectrum to include more bandwidth. This will be necessary as wireless access expands, perhaps matching the widely misinterpreted Moore's Law, suggesting that computing power doubles every 18 months. Voice over IP (VoIP) is already beginning to show up on wireless networks, and this will also grow, especially when it is seen as a less-expensive alternative to land-based phones and can offer instant access to those already logged on for other reasons.

Finally, the emerging 802.16 Wireless Metropolitan Network standard will likely expand across the continent as communities and governments extend the reach to more and more businesses, with smaller wireless networks paying to connect to this service in an effort to expand their reach.

Chapter 2

If You Fail to Plan, You Plan to Fail

• •

In This Chapter

▶ Evaluating your wireless needs

▶ Preparing for a site survey

▶ Doing that site survey

▶ Documenting the site survey

• •

"If you fail to plan, you plan to fail." A simple statement but a profound one. I can't find the source of this quote, but the first time Peter saw it, he was doing work for the U.S. Department of the Navy. Whether you are planning to refit a nuclear submarine or build a wireless network, you must plan to be successful. This chapter sets you up to do a site survey and helps you to plan your wireless network.

It's tempting to skip the planning step and jump right into buying and installing hardware. But you must control yourself. A little planning up front can save you a lot of time and money later on.

Evaluating Your Wireless Needs

To create your shopping list, you must first look at your existing network and evaluate your needs. This step involves asking questions and gathering information. Talk to people about their needs. The more information you gather, the better your plan and ultimately your design. Initially, you will need to answer some very basic questions, such as

  ✔ What is my environment?

  ✔ What is my budget? Or, in other words, how much can I spend?

  ✔ How many clients do I expect?

  ✔ Where will they want to access the network?

  ✔ What types of applications will they use? Or, in other words, what does the data look like?

✔ What technology do I want to use? Or, in other words, what standard do I want to support?

✔ Do I need to protect the data? Do I need to read-protect the data? Do I need to write-protect the data?

✔ What coverage do I need?

The following sections look at these very high-level planning issues one at a time.

What is my environment?

Determining your environment is a logical place to start. Obviously, you need to answer some big questions, such as those regarding location. Is your network indoors or outdoors? The answer to this question might drive all the other decisions. What was used to construct your building? Cement? Metal framing? Is it an office environment? Is it a shop environment with electric motors? Is it a medical environment (a hospital or clinic, for example)? Do you have a cafeteria with a microwave? Do you have an elevator? Do you have wireless mice or keyboards? Do you have a "cube farm"? Do you have office doors? Are they made of metal? Do you have long hallways? A "yes" answer to any one of these questions may cause you problems. For instance, metal walls can *diffract* signals.

Basically, obstacles cause reflections resulting in multiple paths from the source to the receiver, which can have an adverse effect on your wireless network. Wire-mesh is one of the most deadly obstacles; it can scatter almost all your wireless signals. Surfaces such as metal roofs, metal blinds, and metal doors can cause severe reflection and hence multipathing (see Chapter 13).

What is my budget?

Chances are you don't have an unlimited budget (unless you are working on a hush-hush project for the No Such Agency). You have to deal with constraints. The good news is that the price of wireless has dropped remarkably in the last few years.

About eight years ago, Peter co-authored a book on wireless LANs. The technology looked like someone manufactured it in his garage, the data rates were unimpressive, and the standard was awaiting ratification. But the real showstopper was the cost. A wireless bridge cost between $7,500 and $13,000! (All dollar amounts are US.) I bet those babies flew off the shelf. You would expect to pay about $3,500 for a wireless concentrator (a fancy name for what is now called an *access point*). The wireless adapters cost between $425 and $1,500 for 1 Mbps — a real bargain when compared with a $49 10 Mbps Ethernet adapter.

My, times have changed. You can buy an access point for under $25 on eBay. When we wrote this chapter, we found 89 items on eBay, using 802.11 as the search criteria. This included a 2.4–2.485 GHz Tecom +6dB Omni antenna and connectors for a BuyItNow price of $25. (It was at $9.99 with an hour to go.) The starting bid for a new Enterasys 802.11a/b/g wireless PC Card was $63. With less than a day to go, no one had bid on it. It must be overpriced — Peter bought one a year ago for about $119.

We may as well flog this dead horse by providing another example. A year ago a Linksys WAP-11 802.11b Access Point would set you back about $110; now it's available on eBay for $19.99. Okay, so the prices have dropped a great deal. Of course, you will find that 802.11g gear hasn't dropped in price yet, but its list price is greatly influenced by 802.11a and b equipment.

Plan on spending more money than this when building a network for your organization, however. For starters, you should buy equipment that you can upgrade. For instance, had you bought a Cisco Aironet 1200 Series Access Point, you could upgrade from 802.11b to 802.11a and/or g. This device costs around $625. Quite a difference in price. The Cisco device, when compared to Linksys, D-Link, and the other consumer products, looks bad based solely on price.

Cisco designs its products for organizations with larger, faster, and more secure networks, however. Generally, Cisco products have enhanced authentication, encryption, and management functions and interoperate with their other internetworking products. So you really do get what you pay for.

As long as a wireless PC Card is Wi-Fi compliant (see Chapter 1), it should work with any Wi-Fi compliant access point. However, should you want to use the proprietary features such as EAP or longer encryption key lengths offered by a vendor, you may have to buy everything from that one manufacturer. Look around; this marketplace is very competitive at the moment.

Here's one last thought: Vendors offer many proprietary features to try and differentiate themselves from their competitors. But these features aren't for everyone. If you don't need 802.1X integration (see Appendix B), don't pay for it. If you see your solution as having a short payback, you don't necessarily need an upgradeable solution when what you have meets your needs today.

How many clients do I expect?

Obviously, you want to build your network to support the demand within your organization. But does everyone need access today or can you wait and expand it later after you have some experience with radio frequency (RF) technology? Only you or someone in your organization can answer that question. Just don't forget the outsiders.

We travel a great deal and visit many clients. At some locations, we can access their wired network through their access point, and at others, we just can't. Peter just visited a client who had Cisco access points. They were fairly confident that they were secure because (so they bragged) they used LEAP. Peter didn't have the heart to tell them he had ASLEAP (asleap.sourceforge.net) on his laptop (although they did go wild when he connected his laptop to the wired network and got an IP address from the DHCP server). *ASLEAP* is a program you can use to break LEAP and access networks. Other organizations provide wireless access to the Internet to visitors waiting in the lobby. So some organizations plan for outsiders while others don't. But that doesn't mean they won't have outsiders — planned or not.

A fundamental axiom of networks is that they grow. So no matter how much you put in your plan, add some more. Many of us are working in companies that are prospering and growing (while others of us are going through rightsizing — or is it *capsizing*?). We have seen companies with exponential growth. You need to figure out how many clients you will have today as well as next year at this time.

Where will they want to access the network?

If your clients want to use the wireless network only from their desktops, you need only worry about finding a PCI or USB solution. But what if your clients prefer laptops, and they want to access the Internet while having coffee in the cafeteria? What if they want to move from one meeting on the first floor to another meeting on the 22nd floor? This necessitates a PC Card or USB solution but also may involve multiple access points that support roaming. Finally, what if your clients consider wireless networking as the ability to access the organization's e-mail system while waiting for an airplane? Now you need to start thinking about how to accomplish that. Will you use WWAN (Wireless Wide Area Network) and Smart Digital, Compact Flash, PC Cards, or other formats? You should probably give thought to protecting the confidentiality of the data, as well. So it is important to know from where your clients want to access the wireless network.

What does the data look like?

Are people using your wireless network to download Web pages from the Internet? Are they sending graphics? Or are they sending video? Are they playing MUDs (Multiple User Dungeons)? Do they want to use or are they currently using VoIP (Voice over Internet Protocol)? Understanding the data will help you understand the potential load on your wireless networks.

What technology do I want to use?

There is no all-encompassing answer to this question, but here are some scenarios to consider:

✔ **Sharing a broadband Internet connection:** When your primary need is to share a broadband Internet connection, go with 802.11b. Your uplink and downlink capacity will not exceed the 802.11b data rate of 11 Mbps, so it is more than sufficient. In fact, most ISPs provide 2 Mbps or less. Even when uploading or downloading large files, the access point is not the bottleneck; the capacity of your broadband connection is the culprit.

Most Web servers (and especially busy ones) will not serve data any faster than your broadband connection can deliver it. Think of the times you sat there twiddling your thumbs as the graphics. and ads loaded from several different servers in several different locations.

✔ **Moving large files:** If you want to move large data and video files from a client to a server and back, go with 802.11a or g. There is no doubt that 802.11a/g wins hands down when moving files across your intranet. Effectively, 802.11a throughput is 36 Mbps. Granted, this doesn't compare with 100 or 1000 Mbps Ethernet, but it is darned fast. Having said that, it is important to note that unless you are using Giga-Ethernet, your throughput is likely no better than 45 Mbps. You can use 802.11g to stream video without disruption, but be sure to have a policy in place regarding which types of video are appropriate for the office.

In some market segments and applications, 802.11g products will replace 802.11b products, and in others, 802.11b will continue to dominate. The increased throughput for 802.11g comes with a price, which is a required higher signal-to-noise ratio (SNR) that results in a shorter range, higher susceptibility to interference, and a more intensive signal processing that results in higher power consumption. For some applications, such as mobile handsets and PDAs, power consumption will remain a major concern, and these will use 802.11b for a longer period of time. For other markets, such as home networking, 802.11g products will probably replace 802.11b products due to the increased throughput.

✔ **Servicing a large coverage area:** If you need a service coverage area of greater than 80 feet in any direction, consider 802.11b. For every interior wall (made of drywall) that your network must pass through, subtract 20 feet from the product specifications. For any exterior wall or wall of solid construction, subtract 40 feet. The big disadvantage (other than cost) of 802.11a is its range at high data rates. The high data rates drop like a lead balloon as you move away from the access point. So sit on the access point and you'll get 54 Mbps capacity, but don't move too far away. Moving as little as 30 feet or moving to the other side of an interior wall drops the capacity (and throughput) by as much as a third.

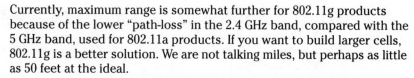

Currently, maximum range is somewhat further for 802.11g products because of the lower "path-loss" in the 2.4 GHz band, compared with the 5 GHz band, used for 802.11a products. If you want to build larger cells, 802.11g is a better solution. We are not talking miles, but perhaps as little as 50 feet at the ideal.

The maximum data rate for 802.11a and 802.11g products is for all practical purposes 54 Mbps, using the same Orthogonal Frequency Division Multiplexing (OFDM) modulation. This is definitely an advantage over 802.11b. However, when an 802.11g product is supporting backward compatibility with 802.11b products, the 802.11g network aggregate throughput available for actual data transport will lower substantially. Which brings to mind, if you have an existing 802.11b network and you are looking to add another access point, 802.11g supports 802.11b clients whereas 802.11a doesn't (because it uses a different part of the spectrum).

✔ **Servicing four or more clients:** If you have four or more clients requiring high data rates, use 802.11a. When you need to support many clients, 802.11a products offer more channels than 802.11b or g products and have the potential to offer more capacity. The theoretical maximum number of clients per access point and the practical number of clients utilizing any access point are two different things. There is nothing inherent to the 802.11g standard that makes it capable of handling more clients than 802.11b. Theoretically, the access point implementation dictates the number of clients capable of sharing its bandwidth. However, because 802.11g offers higher throughput than 802.11b, it can in practice serve the same per-user bandwidth to more clients.

Remember that clients share the access point. If you want to send a file that it is 1MB, it will go faster at 54 Mbps than at 11 Mbps — this only makes sense. You can change the whole equation by adding additional access points to give more clients access. 802.11a definitely has the edge here because you can collocate eight access points in the same physical area, and each can provide 54 Mbps capacity. With 802.11b or g, you are limited to three collocated access points, with each providing either 11 Mbps or 54 Mbps per channel. Earlier in this chapter, we mentioned that the Cisco 1200 Series Access Point supports 802.11a and g concurrently. This is the best solution when you have many clients requiring high data rates because it gives you 11 non-overlapping channels at 54 Mbps per channel. But this obviously costs more money!

802.11b is slightly more robust than 802.11g against in-band interference due to the signal-to-noise characteristics of Complementary Code Keying (CCK) and OFDM. However, implementation plays an important role. For example, certain types and levels of interference will affect 802.11b products as well as 802.11g products, while other types and levels of interference will not affect either, but also a small range of interference levels will affect only 802.11g products. Because 802.11a uses another part of the spectrum, it is not subject to the same types of interference as 802.11b/g.

When money is a big issue, go with 802.11b. You can pick this equipment up for a song (picture the Visa commercial where they sing for their supper). 802.11a is still pricey even when compared with the newer 802.11g gear.

✔ **Providing top-notch security:** The older, cheaper devices tend to have fewer security features. 802.11b equipment typically has shorter keys or only supports Wireless Equivalent Privacy (WEP). It is unlikely that it supports Wi-Fi Protected Access (WPA), Advanced Encryption Standard (AES), or Extensible Authentication Protocol (EAP). You get what you pay for in this life.

When you are setting up a small office/home office (SOHO) network, select 802.11b. It provides enough capacity, is economical, and is easy to install. Otherwise, it may come down to interference (generally 802.11a wins), channels (generally 802.11a wins), coverage (generally 802.11g wins), or cost (generally 802.11g wins).

If you are looking at a Wireless Personal Area Network (WPAN, see Chapter 1), think 802.15. If you are thinking wide area, think 802.16. It pays to adopt widely used standards.

Do I need to protect the data?

Of course you need to protect your data. You should have some knowledge of the type of data you have and its use. How much you spend on protecting the data depends on the data's value. A big problem in most wireless implementations is that people don't use the features built into the products they buy. Peter went network stumbling (read about network stumbling, in which you use software to help you find wireless networks, later in this chapter and again in Chapter 16) in a one-mile radius of his home. Before he left his driveway, he had "discovered" 15 wireless networks. By the time he returned, he had found over 300. Of those, he discovered that about half did not use encryption. To make matters worse, around half of them used the default SSID or network name (see Chapter 3) as well. Our experience shows that this is about par for the course, which is surprising when you consider all the press about the security (or lack thereof) of wireless networks.

What coverage do I need?

Are you trying to provide coverage to a particular location exclusively, such as the boardroom? Or do you have clients who intend to use the wireless network anywhere and everywhere? Do you want to provide coverage outside? Do you need to cover offices with a long common hallway? The answers to these questions affect the number of access points, the power level of the access points, the length of cables, and the type of antennae you use.

If you decide after examining the points covered thus far in this chapter that networking is a fit and you want to do some further investigation, you need to do a site survey.

Preparing for a Site Survey

The *site survey* provides a process for gathering and analyzing answers to the high-level questions covered previously in this chapter. A site survey is your road map to the successful implementation of your wireless network.

The site survey is a step-by-step process whereby the surveyor discovers the RF behavior, coverage, and interference and uses the information to determine the proper placement for hardware. The main objective of the site survey is to ensure that your wireless clients get a strong signal as they use the wireless network — whether they are sitting in their cube or moving about the office. To achieve this objective, the surveyor must analyze the site and discover what coverage is needed to meet the business needs of your clients. Gathering data is key. After gathering the information, the surveyor must analyze the data to glean all possible useful information before acquiring, installing, and configuring network equipment.

You will attempt to define the contours of the RF coverage from the RF source (your access point or bridge). We already mentioned several factors that can affect your coverage in this chapter, but there are many more.

In your organization, it is quite possible that the same individual will do the survey and install the gear as well, but this doesn't mean that you don't need to document your work. You need documentation the next time you want to make changes, or maybe you did such a good job that your boss wants to promote you and you need to hand the documentation to your successor. Regardless, it is important to document your work. When you run into trouble, you may need to walk through the implementation with an expert who will ask to see your site survey straightaway.

Your site survey drills down on the high-level questions covered in the beginning of this chapter. Specifically, your questions will focus on:

- Facilities analysis
- Existing networks analysis
- Area coverage
- Purpose and business requirements
- Bandwidth and roaming requirements
- Available resources
- Security needs analysis

Analyzing your facility

We cover the basics of facility analysis in the section, "What is my environment," but you need detailed information on the facility itself. You can use narrative, photos, video, or blueprints to document the facility. In the narrative, you should spend considerable time describing the type of facility.

For example, if you are a medical facility, then security is especially important (if for no other reason than that the Health Insurance Portability and Accountability Act [HIPAA] makes it so), and you will need to focus on the security analysis.

Hospitals also have wire-mesh glass windows in doors, radiology equipment, elevators, fire doors, long hallways, nurses and doctors on the go, X-ray labs with lead-lined walls, and plenty of government regulations and laws. These factors start you thinking about a potential solution. You need to run the signal down long halls, but keep the signal within the hospital. You also want to consider that you will get RF blockage from elevators and possible RF interference from it and other devices. The lead-lined walls of the X-ray labs will stop signals dead. But the clients are going to roam throughout the hospital. Also, the government only recommends the 2.4 GHz unlicensed band for hospitals. Compare and contrast this environment to a standard office with an open concept and a couple of dozen clients. You might get by with two centrally located access points and rudimentary security. Roaming is probably not an issue since they will access the network primarily from their desks.

These scenarios are different and require different solutions. Each environment is different, but there are enough common characteristics in the type of environment. So study the facility; it will give you clues as to the ultimate environment.

Working with existing networks

Usually, you have an existing wired infrastructure that you want to extend through the use of a wireless access point. If this is not the case, you are lucky. Unfortunately, most of us are working with an installed base, which complicates matters a bit. If this is the case, the first thing you must do before even looking at your needs is to look at what you have. You need to document your existing networks and infrastructure. Draw yourself a network diagram. Better still, get yourself a tool that will do it for you. If you have a limited budget, then look at snmpwalk (`www.trinux.org`), SNMPUTIL.EXE (`www.microsoft.com`), or Cheops (`www.marko.net/cheops`). We recommend WhatsUp Gold (`www.ipswitch.com/products/network-management.html`). It is reasonably priced and starts at $795. If you have Cisco routers, you may already have it because Cisco re-labels it as CiscoWorks for Small Networks. Also, Solarwinds Standard Edition Version (`www.solarwinds.net`) is a real deal starting at $145.

You should also invest in a good drawing program, such as Microsoft Visio for Windows (`www.microsoft.com/office/visio`) or SmartDraw (`www.smart draw.com`). If you are one of those right-brained people — that is, a Mac OS user — you can use ConceptDraw (`www.conceptdraw.com`) or OmniGraffle (`www.omnigroup.com/applications/omnigraffle`). You'll want to keep your drawings and update them as necessary. You will need other tools, too, and we'll get to them shortly.

Most network discovery and management tools provide additional information that you need to document. You'll want to know the operating systems of all servers and clients. You need to especially know the clients because you will most likely need to install and configure wireless utility software or configure OS software, such as Mac OS X or Windows XP.

It's important that you understand the type of data you have because you need to calculate bandwidth requirements for your clients. If you don't know how to do this, pick up a networking book such as *Network and System Integration For Dummies,* by Michael Bellomo and James Marchetti (Wiley). You might want to get out Ethereal (see Chapter 17) and analyze the traffic because you also need to know what protocols to support.

If you already have a wireless network, whether it's a WPAN or a WLAN, you need to know which channels (if any) and what part of the spectrum are currently used.

In addition, consider any security in the wired or wireless networks that you may need to support, complement, or replace.

In many cases, you are going to connect the wireless access point to an existing wired infrastructure, so you must document all the connection points as well as existing power supplies.

Finally, you should understand the naming conventions in use for devices such as servers, routers, bridges, switches, and access points.

Area coverage

You should know whether the new wireless network is going indoors or outdoors. If you are going to put the access point outdoors, you need to bear in mind the weather in your neck of the woods. Are you in Tornado Alley? Are you in the Eastern Caribbean and subject to frequent hurricanes? Do you live in Montréal and suffer through annual ice storms? If you live in an area of inclement weather, you need to take extra precautions. You might consider a *radome* (housing) to protect your antenna or use a grid antenna to offset the wind loading.

Too small for wireless? Think again!

You might think your organization is too small to develop wireless applications. Wrong! Rovenet (www.rovenet.com) Portable Forms lets you create your own mobile data-collection application quickly and economically. You use any word processing program (like Microsoft Word) to edit a template. Then you upload your Portable Forms template using a browser to their server. Rovenet converts the form you created into a portable data collection application. You run your Portable Forms session by collecting and storing the information on your SmartPhone, PalmOS, Pocket PC, or Blackberry device. After you collect the information, you can securely deliver it to Web pages, e-mails, databases, and even faxes. Rovenet acts as your Active Server Page (ASP) for about $100 per year. So, you can start automating your mobile workers right now with little investment.

Outside equipment is very susceptible to attacks. It's a good place to put an active tap on your organization. So take precautions to physically secure your gear. If nothing else, the gear does have some monetary value. A thief could probably get about $9.99 for your antenna on eBay!

In Appendix C, we discuss Fresnel zone and how to calculate it. The *Fresnel zone* is the area around the visual line-of-sight between the sender and the receiver that radio waves spread out into after they leave the antenna. You need to keep obstructions out of your Fresnel zone or else signal strength will weaken.

When you do your site survey, you may find that you need to put an antenna outside. If so, are there any legal ramifications? Do you need a permit? Do you need to contact the FCC or other governmental agency? You need to notify the FCC before constructing or modifying antennae over 200 feet (61 meters) above ground level (AGL). A 10-story building with a 10-foot antenna probably falls in this category. Also consider whether the roof will support an antenna. Do you require additional expertise, such as a structural engineer to help with an assessment of the roof?

Indoors you need to survey the floor layout, firewalls (actual walls, not the kind you use to protect your network), building structure data, wiring room locations, and other information about the environment.

A number of companies sell RF prediction software. *RF prediction* uses computer simulation to estimate the coverage of your access points and the transmit power of the access point antennae. You modify a graphic of your floor plan to create a map that accurately describes the RF characteristics of your building's walls. Then you start virtually placing access points on the map and generating the graphical coverage plot of the expected average signal strength intensity experienced by users in various building locations. This can save you time but is not as thorough as walking around doing a site survey and testing the real signal strength. If you have the money, check out

- ✔ **Airespace:** www.airespace.com/products/AS_ACS_location_tracking.php
- ✔ **Alcatel:** www.ind.alcatel.com/products/index.cfm?cnt=omnivista_acs_locationtrack
- ✔ **Radioplan:** www.electronicstalk.com/news/rop/rop100.html

Purpose and business requirements

The business aspect of your project is where the rubber meets the road. There is no point in installing an access point when there is no business case for doing so. You need to talk to everybody from the Big Cheese to the "early adopter" walking around with the wireless PDA. Find out what everyone intends to do with the network beforehand. Don't treat this as a case of "build it and they will come" (with utmost respect to W. P. Kinsella). Doing a thorough job of documenting your organization's needs allows you to design a wireless network that suits the needs of the organization, as opposed to your personal needs.

If there isn't a business case to implement wireless networks, all you are doing is increasing the costs to your organization. Sure, wireless is sweet, but it better solve a problem, such as wire replacement. In Europe, there are some lovely old banking halls with marble walls and counters. It seems a shame to drill holes in the marble to string cable. Obviously, wireless has an advantage over wired when we are building temporary networks, such as at a tradeshow. And using wireless networks is preferable to laying cable on the floor or stringing it overhead in a warehouse.

Business has typically underspent in supporting mobile users, but there are many examples of business cases for wireless — you just need to find the right one for your organization. For example, wireless WANs are becoming popular in the real-estate market. Agents use wireless PDAs to download maps and information about all the properties in the neighborhoods that pique their clients' interests. As another example, technical support staffs in many industries, such as telecommunications, are downloading problem tickets in real-time to their PDA or handheld. While visiting customers, they can create new tickets for additional work without having to go back to the office. This saves the technician having to return at another scheduled time and provides the customer with a higher level of service. It's a win-win situation.

If your mind needs stimulation coming up with an application for wireless in your business, start with the case studies at www.torwug.org/CaseStudies/main.asp or www.mob1le.com/case.html.

Bandwidth and roaming requirements

Your bandwidth and roaming analysis might actually determine the type of technology you purchase and use. If you find that your clients intend to use the network to scan data in the warehouse and send the data to the central server, the bandwidth requirements are low. That scanning device probably needs only 2 Mbps bandwidth, but clients need seamless connectivity when moving about the warehouse. On the other hand, the clients might design car parts and need to upload and download technical diagrams on a consistent basis. These clients need as much bandwidth as they can get. These are your bandwidth hogs. (Picture a pig going through a snake.)

As part of your bandwidth analysis, you must understand how many clients will access the network from a particular area, such as the boardroom. You must understand that the needs in your organization are not uniform across the organization. That is, one group might use the wireless network more than another. A group's bandwidth requirements typically result from the types of applications they use. Do they send time-sensitive data or not? Do they use connection-oriented applications?

Your clients might want to roam indoors, outdoors, down the street, and across the country. Roaming is not trivial as we cross boundaries from one interconnectivity device to another. Maintaining VPNs (see Chapter 12) is especially problematic with roaming. You need to capture these requirements upfront before selecting software or hardware solutions. You may find that there are areas requiring special connectivity solutions, but you won't figure this out until you know your clients, their applications, their data, and their bandwidth requirements.

Available resources

We talk about budget concerns in the earlier section, "What is my budget?". How much money do you have? How much time? These are really functions of the same thing. You can buy time with money and vice versa. But you must also determine whether you have the human resources to design, implement, and support wireless networks. Do you have people capable of training your clients on the use of the wireless networks and applications? Are the resources available to implement and support the various components? Or do you need external support?

Security needs analysis

Last, but surely not least, is the need to do a security needs analysis. You need to document the threats to your wireless network and the possible threat agents. Examples of threat agents include your competitors and disgruntled ex-employees. You need to assess the likelihood (the risk) that the event will happen. Then you need to calculate the impact on your organization (your vulnerability) should the event occur. Many good books are available that cover security, but you may want to start by looking at *Network Security For Dummies*. If you want to test your network, check out *Hacking For Dummies*. And, if you give up right now, check out *Golf For Dummies;* you will soon have lots of free time on your hands. (All of these books are published by Wiley.) If you haven't given up, we cover security in Part III.

Successful security programs are built on solid foundations. You must develop or adhere to any organizational security policies, standards, and guidelines in your organizations. If you don't have them, this is the time to start working on them. It will take time to get them approved by the powers that be.

Your site survey must cover a security section detailing the level of security required. When drafting the report, you must consider the security posture of your organization, the nature of the data, and the knowledge of your clients.

Developing a site survey checklist

We don't necessarily support checklists as a way of doing business. Experience counts — and you can count on experience. But a checklist is an excellent *aide-mémoire* or tickler. Here is a start on your checklist. Add and remove items as you see fit: Make it your own.

Site Survey Checklist

- ❑ Organizational policies and standards
- ❑ City/town/county ordinances
- ❑ FCC regulations or the equivalent for your jurisdiction
- ❑ Budget
- ❑ Building plans or blueprints
- ❑ Power diagrams and information
- ❑ Current network topology, layout, diagram or schematic (call it what you will)

❏ Interviews notes:

    ❏ Clients

    ❏ Network Administrator

    ❏ Security Administrator/Officer

    ❏ Facilities Manager

❏ Remodeling plans (where applicable)

❏ Access to all areas of the facility where wireless is needed

❏ Access to wiring closets

❏ Access to roof (for outdoor antenna installation)

❏ Site survey gear

Using Site Survey Equipment to Get It Right

We discuss some of the tools you'll need, such as drawing and network discovery software, earlier in this chapter, but there are more. You need to amass some tools to do the physical part of the site survey. This is where you get your hands dirty.

Your basic gear includes:

- An access point or bridge
- Various antennae
- Antenna cables and connectors
- Laptop computer or PDA
- Wireless adapter (PC Card, Smart Digital or Compact Flash) with latest drivers and utility software
- Site survey software
- Double-sided tape
- DC/AC converter and batteries
- Digital camera for taking shots of the various locations in the facility

If you can find an access point with variable power output, use it. Variable power allows you to try some cell shaping and sizing. You'll want to use double-sided tape to attach the access point to various locations in your organization to measure the signal strength. Because you may not have power everywhere you need to put an access point, or because it may be too difficult for your survey to plug into the mains, it helps to have an access point and a DC/AC converter. You can plug the access point into the converter and use battery power. Of course, your high touch/low tech people can kludge it by dragging extension cords all over the place!

If you can swing it in your budget, TerraWave (www.terra-wave.com/SiteSurveyKit1200.pdf) offers the whole kit and caboodle. If you don't have $5,349, you'll need to do it yourself. The kit probably pays for itself after a while, but we cannot justify the cost to *our* boss. But take a look at what you get for your money, and you can find what you should start collecting for your own kit. Note the battery pack (and you thought we were kidding). And you don't even get the laptop or site survey software!

Most quality wireless PC Cards come with a site survey or monitor utility. Figure 2-1 shows you an example of a typical one. At a minimum, you need a link speed indicator and signal strength indicator (see Figure 2-2).

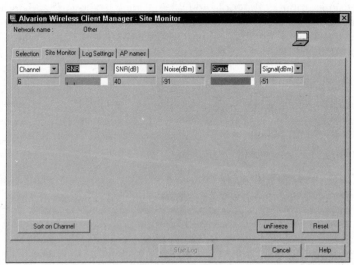

Figure 2-1: Site monitor.

However, to perform a worthwhile site survey, you really need signal strength (measured in decibels relative to 1 milliwatt [dBm]), noise floor (measured in dBm), signal-to-noise ratio or SNR (measured in dB), and link speed. (If you aren't familiar with these terms, jump to Appendix C for a quick start. You need to know these concepts when you calculate your link budget.)

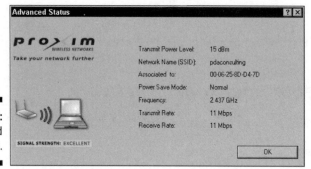

Figure 2-2:
Advanced
status.

If your client utility doesn't do the job, you can use Network Stumbler, that is, NetStumbler (http://www.netstumbler.com) or its kin. (Chapter 17 provides a list of these network discovery tools. We also visit them in Chapter 16.)

To test signal strength, you walk around using your site survey software measuring and marking down signal strengths on your facility's floor plan. Pay attention to the signal-to-noise ration (SNR) calculation because it shows the viability of your RF link. Anything higher than 22 dB is a good link.

If you are planning on several access points and many clients, you may want to invest in a wireless "sniffer" or analyzer. A product like AiroPeek (www.wildpackets.com) provides the data needed for our site survey. We talk about these products and how to use them in Chapter 13. A wireless sniffer provides information about nearby WLANs with respect to channels in use, distance from you, and signal strength.

If you have the budget, you can use a spectrum analyzer, but they are expensive. They are really useful when you have interference and you cannot find it. If you encounter this problem, you may need an expert with the right tools for the job.

Our list is not complete, but should you need anything more than the items on the original checklist, seek professional help. (We mean a *wireless* specialist.)

Doing That Site Survey

After you have the gear, you can do the analysis. An important step in the development of any plan is the gathering of information. If you want to make the best decision, you need the best information. In this section, we show you how to calculate a link budget and find out what information you need to do your planning well.

The information you gather varies on whether you are inside or out. One thing that doesn't change is the requirement to do a link budget. Sooner or later, you need to figure out whether the link is viable.

Analyzing your indoor network

For your indoor survey, collect and record the following:

- ✔ AC power outlets and ground
- ✔ Wired network connectivity points
- ✔ Potential RF obstructions such as fire doors, metal blinds, metal-mesh windows, solid walls, lead-lined walls, and so on
- ✔ Potential RF interference from microwave ovens, elevator motors, 2.4 GHz cordless phones, baby monitors, Bluetooth, and so on
- ✔ Clutter — you know, those places where they pile up listings, boxes, and other junk

Analyzing your outdoor network

For your outdoor survey, collect and record the following:

- ✔ Trees, standing water, buildings, and so on
- ✔ Visual line of sight and RF line of site (see Appendix C)
- ✔ Link distance
- ✔ Weather
- ✔ Antennae and tower accessibility
- ✔ Roof accessibility

Calculating a link budget

You need to figure out whether your proposed link is viable. You can figure out the link budget to determine the viability of the link. When calculating a link budget, you need to either look up or figure out the following values based on your proposed equipment and environment.

- Transmitter output power (dBm)

- Transmitter antenna gain (dBi)

- Transmitter-side cable loss (dB)

- Transmitter-side connector loss (dB)

- Path or free space loss (dB)

- Receiver antenna gain (dBi)

- Receiver-side connector loss (dB)

- Receiver-side cable loss (dB)

Set up a spreadsheet with the preceding list in one column and gains or losses in another column. You now need to fill in the second column. Here is an example to help you:

- **Cisco 1200 Series Access Point:** We discussed this product previously (the product specification is available at `www.cisco.com/warp/public/cc/pd/witc/ao1200ap/prodlit/casap_ds.pdf`). This product comes in 802.11a, 802.11g, and 802.11a/g models.

- **Tecom Omni antenna:** This is a cheap, easy-to-install, and durable antenna made for short-range, point-to-multipoint environments. We intend to propagate the signal only about one-tenth of a mile, just over 600 feet.

- **Coax cables:** For this configuration, you need two lengths of cable and two connectors. The first cable is a 24-inch (2-foot) pigtail made of LMR 240, and the second is a 60-inch (5-foot) pigtail also made of LMR 240 coax. The connectors are N-type connectors. Product information is available at `www.timesmicrowave.com/telecom/lmr/LLPLcat.pdf`. The cable loss is 12.9 for a hundred feet, including the NM connectors.

- **Orinoco 11a/b/g/ ComboCard:** This card supports 802.11a and g. Product information is available at `www.proxim.com/learn/library/datasheets/11abgcombocard.pdf`.

After you know the configuration, you can calculate the link budget as follows:

- **Transmitter output power:** According to the product specification, this again will depend on the standard you choose and the jurisdiction you live in. Assume for this example that you use the maximum OFDM figure of 30 mW or 15 dBm.

- **Transmitter antenna gain:** +6 dBi.

- **Transmitter-side coaxial cable and connector loss:** –0.90 dB.

- **Path or free space loss:** –84.34 dB.

- **Receiver antenna gain:** +6 dBi.

- **Receiver-side coaxial cable and connector loss:** –0.90 dB.

You can find the formulas for these calculations in Appendix C. When you have all the calculations, you simply add the values together. For example, you add the gains together. That's easy:

2×6 dBi, or +12 dBi

Next you add the losses together:

$2 \times (-0.90) + -83.34$, or -85.14 dB

Next you can add the signal strength to find the loss:

Signal strength = 15 dBm + 12 dBi − 83.34 dB = −56.34 dBm

The resulting value is the signal strength at the receiver. But what does that tell you? Sometimes your vendor will publish receiver sensitivity. For example, Cisco does. But sometimes they don't — D-Link doesn't. (You can find receive sensitivity for many common access points and PC Cards at `freenetworks. org/moin/index.cgi/ReceiveSensitivity`.) If you cannot find your receiver's sensitivity, you can use the figure −80 dBm.

Generally, a receiver's signal strength falls between −83 dBm and −90 dBm. (Appendix C discusses dBm, as well.) Our Cisco 1200 access point has a receive sensitivity value between −68 dBm and −95 dBm, depending on the standard and the data rate. So, you add up all the numbers and determine whether the final signal strength falls above or below your Wi-Fi receiver's signal threshold. For instance, should you want to use 802.11a at its highest data rate, you need at least −68 dBm signal strength; and for 802.11g, you need −72 dBm.

When signal strength is above or "in the black," you're okay and can establish a link. Problems occur when you are "in the red." You then have to start looking at your RF circuit. Can you get lower-loss cables? Can you tweak the transmitter (within legal limits, of course)? Can you get a higher-gain antenna? Can you use a wireless repeater? Can you use a transmit or receive amplifier? Can you use a power injector? Somehow, you have to push the numbers back into the acceptable range. But in our example, we are well within the published limits.

When designing your network, you may want to recommend a 20 dB margin to cover reflections and diffractions that could interfere with your link. If you calculate that your signal strength is 20 dB over the receiver sensitivity, you're golden. If you want 54 Mbps 802.11g, you may have a problem because 72.0 − 56.0 = 16 dBm. You may find that some clients are fine while others complain about throughput. You could always give them a higher-gain antenna. For instance, giving the client a +15 dBi antenna instead of the +6 dBi used here would give you another +9 dBi and put you well over the top.

One last word: If you are math-phobic, you'll need help doing the link budget, but you must do it. You don't want to recommend to management (or yourself) a configuration only to find out that the link is not viable and you need to upgrade the cable, connectors, antennae, access points, or wireless clients. It's embarrassing to have to approach management (or your spouse) to sheepishly ask for more money because you blew it. Remember, budgeting is done as a planning exercise. You then compare actuals to budgeted estimates and figure out what when wrong, which is covered in Part IV.

Describing Your Final Plan in a Site Survey Report

After you finish the analysis, it's time to write up your findings. The report you create serves as your roadmap for making the final equipment selection or for preparing a Request for Proposal for external help. You also will want to keep the report as permanent documentation so that others can see your hard work.

Unlike SEC filings, there is no law or regulation specifying the format or nature of the reporting. But you will want to get your point across. You most likely will use the report to justify expenditure on software and hardware, so you must make it comprehensive and accurate.

Defining the business requirements and methodology

Restating the business requirements (that is, wants and needs) for any IT project (or network) is always a good place to start. If a misunderstanding or misinterpretation of the business needs exists, it's a good thing to get them out of the way straight off. Put this section up front and get your organization's agreement.

You are going to make some conclusions, such as the type of network and standard to choose, so you need to define for the reader the methodology you used. (We hope you use this book!) The reader can then look at your process, decide whether it is a good process, and decide whether you came to the right decision based on the work done.

Documenting the requirements

After the background information, you must tell the reader the following:

- Hardware choices for access points, bridges, repeaters, amplifiers, clients, antennae, and attenuators
- Cables and connectors
- Hardware placement, including antennae, and configuration information
- Access point and antenna mounting instructions
- Power supply and power settings
- Wireless network coverage using diagrams to get your point across
- Anticipated throughput for the coverage area, which you figured out from the use of your site monitor utility, network discovery software, or wireless sniffer
- Potential or actual sources of interference, which is information you captured from the wireless sniffer or spectrum analyzer
- Other problem areas, such as potential wired or wireless networking problems
- Security solutions and configuration
- Estimate of costs
- Proposed implementation work plan

After reading your report, management should have no doubt that you did a thorough analysis and they should understand the required action. Yes, planning is a tough job, but that is why planners get the big bucks. Yes, planning requires a lot of work. But, doing the work on the front end will save you a lot of time, money, and grief on the back end. After you finish your plan, you should find it a relatively easy job to match technologies to your plan, as covered in Chapter 3.

Chapter 3

Matching Wireless Technologies to Your Plan

*T*o ensure that your wireless dreams come true, you need to follow a plan. Luckily, Chapter 2 is all about planning. This chapter shows you how to choose and use the right equipment to go with all that planning.

Choosing the Right Networking Hardware

A plethora of wireless equipment is available to you. Deciding which to use is part of the battle for wireless supremacy and nights with sound sleep. Before you can determine the right hardware to use, though, you need to understand the different wireless networks and ensure that you are building the one you want and need.

Although you may understand and recognize the standard infrastructure type of network, in some situations, you may choose a peer-to-peer network. In a *peer-to-peer* network, each machine uses its network card to connect to another machine's network card. No wires, of course, and no central server, or "boss." You see examples where peer-to-peer networks are used in Chapter 8, along with examples of how this type of network might evolve as time progresses. A peer-to-peer network works fine for a small number of devices and is often used between Bluetooth-enabled PDAs or laptops, for example, as well as between laptops or computers with wireless network cards. This type of network doesn't usually provide a connection to a wired local area network (LAN) although if one of the computers is set up to do the routing, it is possible.

The other more familiar network involves using some form of access point or base station to achieve a network more similar to your business's local area network. This is typically an extension of your LAN, with the wireless connectivity adding to the already well-established wired connections. You can see the differences in Figure 3-1.

Peer-to-peer

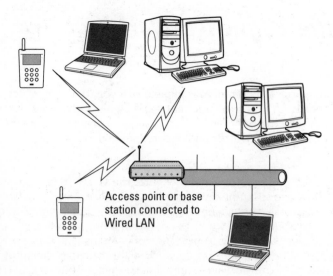

Access point or base station connected to Wired LAN

Figure 3-1:
Examples of
wireless
networks.

So you determine that you require one or both of these wireless networks. Now the work begins. Which hardware do you choose? Is one better than another? Can different vendor equipment co-exist?

Most wireless networking hardware vendors support the 802.11 standard, so they can theoretically interoperate. However, we recommend that you verify whether the equipment that you purchase interoperates as the standard is a fairly recent one. Also, note that vendor implementations are not always ideal. For details on whether the devices you purchase will interoperate, you can go to the Wi-Fi Alliance (www.wi-fi.org/OpenSection/certified_ products.asp?TID=2) and check this site for certified products and their degree of interoperability. It is often just easier, though, to purchase the same vendor equipment rather than suffer the potential of operability problems. Additionally, of course, there are the different methods for wireless communications; Frequency Hopping (FH), Direct Sequence Spread Spectrum (DSSS or DS), and Orthogonal Frequency Division Multiplexing (OFDM), which are not interoperable, so your equipment needs to use the same frequencies everywhere. We explain a little bit of how they differ later in this chapter. In case you're interested, Appendix C provides excruciating detail about these methods.

If you insist on purchasing wireless networking hardware from separate vendors, be sure to obtain guarantees from them that if the hardware does not interoperate or follow the standards, you can return it for a refund. Naturally, there are differing 802.11 implementations, and you need to be aware of each because they are not all compatible.

You may decide to use Cisco products for your network and if so, you certainly can't go wrong. You may find less expensive products from other vendors, but as with all things, you need to shop with a list of criteria that are appropriate for your needs. Some of the things on that list might include

✔ Type of device needed (access point, repeater, bridge)

✔ Your budget

✔ Vendor implementation assistance

✔ Vendor technical support

✔ Support Availability (7/24/365)

✔ Centralized management options to ease configuration

✔ Reporting capabilities

In Appendix B, we tell you all about the different wireless standards such as 802.11a and 802.11g and which interacts with which. Read that Appendix before you go shopping to ensure that you buy devices that will interact.

Are You Being Served? IBSS, BSS, and ESS

This section is fairly technical and is most useful to those of you performing wireless network maintenance, so feel free to skip over it and continue on to the next section.

A collection of wireless cards (or access points) that are within proximity of each other and are communicating together is known as a *Basic Service Set* (BSS). When stations are participating in this BSS, they share certain common network parameters. For example, they all transmit and receive on the same channel, and they use a common BSSID. This information is broadcast in what are called *beacon frames* that are sent out at regular intervals.

Because some of these parameters can be similar, such as the same channel, or common data rates, a unique identifier is necessary. That identifier is a 6-byte number that is used to identify the Basic Service Set and is referred to as the *BSSID*. By sending this identifier in each packet, the network is assured that all devices are supposed to be communicating. This is not a security function, however, and offers no protection against unauthorized access. It merely ensures that devices can communicate with each other.

The BSSID is not the same as the SSID. The Service Set Identifier (SSID) is the 32-byte maximum string commonly called the network name that identifies which ESS (Extended Service Set) or IBSS (Independent Basic Service Set) to join. ESS and IBSS are explained later in this section.

Two kinds of BSS identifiers exist. One is the independent BSS (IBSS), and the other is the infrastructure BSS. An IBSS usually refers to an ad hoc network meant for peer-to-peer networking such as what can happen when numerous people get together and use their devices to communicate or perform some simple file sharing. All the devices can hear each other, and packets are directed to the recipient device with all devices sending out beacons with a randomly generated BSSID. You see how this type of network offers new chances for gatherings in Chapter 8.

In an infrastructure BSS, there is at least one access point rather than a collection of peers. When a device wants to send data to another device, the packet is sent to the access point that then forwards it to the expected recipient device. This setup uses the MAC (Media Access Control) address of the access point as the BSS, which is the only one sending out beacons. It is sometimes referred to as the *BSS master* with the client devices called the *BSS clients*. As you can see, there is a little less traffic in an infrastructure setup and more organization with only the one primary device. This is how most of our wireless networks function.

802.11 networks therefore grow as needed by combining these infrastructure BSS's into larger networks called *Extended Service Sets* (ESS). In order for devices on one BSS to talk to others in another BSS, they need a special service. This is where a distribution system service or DSS is used. This service connects all the relevant BSS's into one big happy family. Whew! Confused yet? Knowing how all this works is useful when things go wrong so hang in there, there's more to come. Figure 3-2 shows how all these fit together.

When you connect a laptop or PDA to a wireless network, you usually set the Service Set Identifier (SSID) on your network card, which is a maximum 32-byte string that identifies your IBSS or ESS, or more typically, your single infrastructure BSS. This is often called the *network name,* and it's how your device identifies which network to join.

Using the same SSID in two or more infrastructure BSS's allows the device to roam between those different networks because they both use the SSID to designate they belong to an ESS. You can find more on roaming in Chapter 7.

Figure 3-2:
Example of
BSS, DSS,
and ESS
interaction.

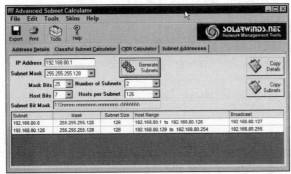

Selecting the Wireless Mode

We discuss modes of operation a little in the preceding section, referring to them with all kinds of acronyms. Most organizations are interested in using an infrastructure network, which they set up and run like a wired local area network.

Considering ad hoc mode

Ad hoc. Sort of evokes the idea of random or thrown together, doesn't it? Well, in a way, that is what an ad hoc network consists of in the wireless world. Technically, it uses an Independent Basic Service Set (IBSS) and has no capability to grow beyond adding the number of devices in range. It is great for the coffee shop or lunchroom. Whom you can communicate with in an ad hoc network is limited by the range of your device; there is no central access point to blast out waves to let you log in and use the network. You can either reach the PDA near you, or you are too far away. It's that simple.

Use this mode for community group meetings where everyone carries some form of wireless device, such as a laptop, a PDA, or a tablet PC. It is great for informal chats and small data transfers, such as business cards or that pretty girl's home phone number. One other good use for an ad hoc network is to connect with someone else in the meeting you are both attending and share comments or browse each other files — anything to relieve the boredom of being in the meeting. Of course, if you are fired for inattention or dereliction of duty or something, don't blame us.

Using infrastructure mode

Infrastructure networks are the typical wireless networks you all know and hopefully love. Coverage is determined by the range of the access point. Coupled with a good external high gain antenna this can be quite extensive. When you add in all the other possible options, like bridges and repeaters, it gets downright giddy as to how far you can travel while remaining connected.

You can use this to extend your wired LAN or an Internet connection or even to add a wireless printer to your network. The possibilities seem endless. You need an access point and to assign the SSID to those devices connecting to your network. Most of this book covers this type of wireless network because it is likely the most useful or certainly the most common at this point for business needs.

Along with the potential for distance, this type of structure also offers increased security options such as WPA and EAP, which are covered in Chapter 11.

Gearing Up to Send and Receive Signals

Okay, so you want to set up your wireless network. Before you do so, you need to understand a little about how signals occur in your network so you can ensure that the users get logged on and can perform their business.

Frequencies and spectrums

No, we are not talking about how often it occurs. It's a different type of frequency. And there is a broad spectrum of results you can get from that. Now that we are finished with the word games, let's talk about what we mean.

For the very technical among you, Appendix C offers a huge amount of technical detail you can ponder, so search out that Appendix for more detail. Here, we cover the important parts so everyone is working on the same page.

Familiarizing yourself with frequencies may be daunting, but consider that your wireless network is simply a radio station and needs to operate within a set of parameters so that it doesn't conflict with other radio stations. You might listen to WKRP in Cincinnati (did that actually exist?) or CHFI in Toronto. If so, you turn your radio dial to the frequency they operate on, such as FM 98.1 or 99.9.

Radio frequencies are high frequency alternating current (AC) signals passed along copper wire or conductor until an antenna radiates them into the air. Signals are technically identified by cycles per second (Hz). The Hz stands for *hertz* or vibrations per second, with kHz (kilohertz) standing for thousands of cycles per second, and MHz (megahertz) meaning millions of cycles per second. Wow. Commercial FM radio uses signals between 88 MHz and 108 MHz, or 88 million–108 million cycles per second. There are a number of these bands that are detailed more explicitly in Appendix C.

Wireless access points need to do the same thing — operate where they do not infringe on other uses. This is where frequencies come into play for the wireless world. The various 802.11 frequencies operate at 2.4 GHz and at 5.8 GHz. 802.11b wireless devices are generally effective up to about 300 feet using the 2.4GHz band. 802.11a coverage can often meet acceptable reliability at over 200 feet with a faster throughput than the 802.11b frequency. Products based on 802.11a use the 5.8 GHz band. The other popular wireless range is 802.11g, which runs in the 2.4 GHz range and offers backwards-compatibility with 802.11b devices. 802.11a operates on a different frequency than 802.11b/g, which is why 802.11g is backward-compatible with 802.11b, and 802.11a is not. So if you have a big investment in 802.11b, you may want to go to 802.11g rather than switch to 802.11a.

Spectrum is the term used for the range of electromagnetic radiation from the highest to the lowest frequency. Everything from x-rays to visible light and radios use some part of the spectrum although our wireless frequencies generally fall into the low-range frequencies, from 2 kHz to 300 GHz. Go any higher, and you hit infrared and gamma rays. Nobody has figured out how to use the gamma ray frequency for communication (that we know of), so we don't need to worry about understanding that level of frequency.

Get the right antennae

A serious wireless network involves the use of external high gain antennae. It just enhances the capability of getting a good signal and allowing your users to connect wherever you have designated that they should. Antennae do not provide more power: They are merely directional devices that channel the power produced in your access point to farther locations. A good analogy might be the cone megaphones that cheerleaders use. These work like an antenna to focus the cheerleaders voice in a particular direction so that you can hear them better. Wireless antennae do the same thing.

There are many types of antenna available. You need to analyze the differences and decide which might be best for you in your wireless environment. Table 3-1 shows some of the different types and describes their use.

Table 3-1	Types of Antennae
Type	*Description*
Patch antennae	These antennae consist of a square conductor mounted over a groundplane. These are directional antennae.
Omni-directional antennae	These antennae provide a 360-degree radiation pattern. They are used when coverage in all directions is needed.
Parabolic (dish) antennae	These antennae focus the signal in the direction they are pointing. They are typically used in point-to-point networks between buildings, such as in television and satellite communications.
Rod antennae	These antennae radiate a signal 360 degrees. Locating it horizontally or vertically impacts the range it produces. The longer the rod, the better the beam. These antennae are typical in home or small office wireless devices.
Sectorized antennae	These antennae are similar to rod antennae but are able to transmit in sections, such as only 90 degrees rather than the full 360. Like the rod, they are used in multi-point solutions.
Yagi antennae	Yagi is a type of directional antenna popularized by CB and amateur radio aficionados.

There are many technical discussions concerning types of antenna. Table 3-1 offers only the more common ones to reduce space. In commercial applications, you need to define your requirements and purchase the antenna that fits your needs. You might already use a dish antenna for your satellite television and know they need to be focused just so in order to obtain the best signal. Wireless antennae are little different. They require proper alignment and focus in order to maximize their range.

Figure 3-3 shows the different types of antennae.

We use an inexpensive high gain antenna that we bought on the Internet for war driving. As an alternative, you can build a wireless antenna using a Pringles can. Tasty little things, those Pringles, and the can is useful, as well. The Pringles antenna is generally considered to be a Yagi-type device. An Internet search on Pringles antenna yields hundreds of results, so feel free if you want to find out how to build one for your personal use. You can find some pictures of Yagi antennae at `http://seattlewireless.net/index.cgi/DirectionalYagi`.

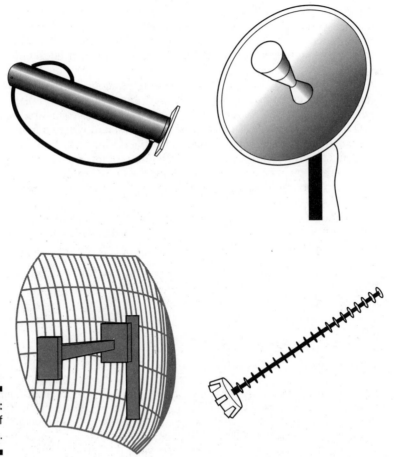

Figure 3-3:
Examples of
antennae.

So which do you use? That's a little harder to answer and often needs expert advice based upon your specific needs. Here are some general guidelines:

- ✔ Use a **parabolic** or **dish antenna** in point-to-point situations where you need the greatest range.

- ✔ Use a **rod antenna** for general purpose or mobile needs where longer length can be obtained if needed for greater range.

- ✔ **Sectorized antennae** are used for point-to-point installations on customer devices, as are **Yagi antennae.**

Introducing the zone — a wireless diet

Are you in the zone? Working at it? That's great because now we have another zone for you to follow. Using wireless equipment allows a level of mobility never experienced before, and as you visit lunchrooms, coffee shops, libraries, university campuses, and other locations, you are starting to enter the zone.

What is the zone? A location that offers wireless connectivity. A simple concept yet one that has profound implications for how we work and play. Where we were once restricted to our desks, we are now free to roam around and connect in many places. IBM has a great advertisement where a man is standing like a statue in a case, and a group of people is staring at him. The voiceover is saying he is an obsolete executive chained to his desk. It's a good analogy of yesterday's workplace compared with the one of tomorrow as businesses start to introduce the zone to places where they need connectivity.

If you search the Internet for the word *zone* along with the word wireless, you find hundreds of places advertising their wireless zones, many of them universities. Some of these include

- ✔ www.mcgill.ca/ncs/access/wireless/support/access
- ✔ www.inet.co.th/services/wirelesszone
- ✔ standards.ieee.org/wireless
- ✔ www.lib.uiowa.edu/hardin/wireless.html
- ✔ www.ezgoal.com/hotspots/wireless/score.asp?FileID=86181

If you are using your WAP-enabled phone, you can go to a Wi-Fi Zone Map and locate nearby wireless zones using wap.wi-fizone.org on your phone's Web browser. Selecting this option takes you to menus for country and state and then city locations. If it's games you want for your device, go to the wireless zone at www.gamespot.com/company/bluelavawireless.html and browse away to your heart's content.

Understanding and Using Layer 2 and 3 Concepts

Layers? Now we are talking about making a cake or something? No, it refers to the Open Systems Interconnection (OSI) model for managing network traffic. The OSI model refers to seven layers, as shown in Table 3-2.

Table 3-2		OSI Layers
Layer	**Description**	**Use**
7	Application	This supports application and end user processes.
6	Presentation	This is where data is transformed into a form the application layer can accept.
5	Session	This layer manages user sessions and dialogues and controls establishment and termination of logic links between users.
4	Transport	This provides transparent transfer of data and is responsible for error recovery and flow control.
3	Network	This layer provides switching and routing technologies, enabling your data packets to traverse the network.
2	Data Link	This is the layer that takes your data and encodes or decodes it into bits (zeroes and ones). It uses the MAC address in determining the proper machine to forward the data to or from.
1	Physical	This is the layer that actually carries your data across the network at the electrical and mechanical level.

All this is very nice, and everyone talks about the OSI layers, but the reality is that they don't translate well into the TCP/IP world because one came before the other. Nevertheless, the general approach is still followed, and this is where the term Layer 2 and Layer 3 originate. As your data passes from one machine to another, it travels up and down this protocol stack, and you are presented with the results. For example, when your browser sends an HTTP request to another machine, the request travels down the layers from 7 to 1 and across the network and back up the layers until your request is fulfilled. Without getting too detailed, how far up and down that protocol stack the request travels depends on what you are doing.

As you can see from Table 3-2, Layer 2 is the Data Link layer and Layer 3 is the Network layer. How does this matter to our wireless network you ask? So glad you did. The IP address that you use runs at Layer 3 while your network card's MAC address is used in the lower Layer 2.

WEP (Wired Equivalent Privacy) and its incarnations operate at Layer 2 on a wireless network. If you turn on MAC addressing restriction on your wireless access point, this uses Layer 2 because that is where the MAC address is "understood" by the network stack. When you use a virtual private network (VPN), that primarily uses Layer 3 although the client and server ends typically go up the entire stack, and a few vendor solutions use Layer 3. 802.1x is a new protocol standard that defines Layer 2 authentication for both wired and wireless networks and allows for dynamic port authentication and authorization. Basically, this means you need to authenticate to the network in order for your device to connect to a switch or in our case an access point. This is a good start at preventing unauthorized users on a network. Of course, full encryption is really the only way to completely protect the network, but that is years away for most companies.

Using encryption like a commercial VPN or Windows IPSec protects the data on the encrypted portion of the network from Layer 3 down Layer 2 and 1 and across the network to the other machine. You can read a good article on it at www.interop.com/lasvegas2004/pdf/which_layer.pdf where it goes a little further in-depth than we can here. Companies like 3Com provide wireless devices that provide Layer 3 encryption such as their 3Com AirConnect 11 Mbps Wireless LAN with added VPN tunneling built into it. If you ran an application that offered encryption, such as PGP (Pretty Good Privacy), that operates at Layer 7 and therefore adds some overhead to the network (because all the packets are encrypted at that layer and pass across the network that way). Encryption generally adds to the size of each packet and therefore there is a cost in speed. So the lower it can be done on the protocol stack, the better. Hardware-based encryption operates at these low levels; this, along with the fact that dedicated equipment is being used, typically makes it faster than software-based solutions.

All this is well and fine, but deciding which to use is still an exercise in judgment. Because WEP or WPA runs at the lower layer, they can be generally thought to offer better performance and can be used along with other security like your VPN to improve the overall security posture of your wireless network. However, they suffer from weaknesses in implementation and are not the industrial strength needed by corporations. They offer no individual authentication, and managing the keys is somewhat cumbersome although newer management solutions are being introduced almost daily, it seems, so this situation may change. For now, a secure environment is going to use both the Layer 2 wireless security protocols of 802.11x or at least WPA, as well as those Layer 3 components like VPNs.

Part II

Implementing Your Wireless Network

The 5th Wave   By Rich Tennant

©RICHTENNANT

"I guess you could say this is the hub of our network."

In this part . . .

In this part, you get serious about using a wireless network and discover how to quickly start with IrDA and
Bluetooth. You see how to connect your office printers
and keyboards through a wireless connection. Then you
install and configure your access point and connect your
client machines to the wireless network. The part ends by
addressing the connectivity issues you may face as you
roam around. It even tells you how to remain connected
to the office as you travel through airports and hotels.
Your journey takes you from the WPAN to the WLAN and
finally to the WWAN.

Chapter 4

Getting a Quick Start with Wireless Personal Area Networks

*W*ireless technology is not new. Over a hundred years ago, Guglielmo Marconi stood on Telegraph Hill in Newfoundland and experimented with wireless telegraphy. We have come a long way in 100 years, and perhaps even further in the last 5 years. Portable and mobile computing use is growing rapidly in the 21st century. Every company recognizes that to compete in the global market, they must deploy mobility solutions. Mobility is what the IrDA (Infrared Data Association) standard and Bluetooth provides. In fact, mobile computing has grown dramatically over the past few years as a result of IrDA and Bluetooth.

Although the IrDA protocol has been languishing in the last few years because of the emergence of the more efficient and higher capacity Bluetooth protocol, you cannot overlook its importance as a pacesetter for Bluetooth. Bluetooth, in turn, may or may not lose out to an emerging technology.

Understanding IrDA

Infrared, although not generally used for WLANs, was part of the original 802.11 standard. Normally, you use infrared for proximate or personal networking and not local area networking. In 1993, the leaders of the communications and computer industry came together to form the *Infra-red Data Association* (IrDA) (www.irda.org) with the purpose of creating a standard for infrared wireless data transfer. They developed the IrDA Standard to facilitate inexpensive point-to-point communication between electronic devices (for example, computers, mobile phones, and peripherals) using direct beam infrared communication links through free space. IrDA's strength is its versatility. Look around your office, and you will see infrared used on many different devices. You might find it in your laptop or the remote control for your PowerPoint presentation.

IrDA has two standards: IrDA-Control and IrDA-Data. *IrDA-Control* is a low-speed protocol for wireless control devices such as mice, joysticks, and remote controls. There are many protocols within the IrDA-Data standard. One protocol ensures that IrDA devices don't fight among themselves during multi-device communication. There is only one primary device, and others are secondary. Also, another protocol describes how the devices establish a connection and close it, and also how they are internally numbered. As soon as information about supported speeds is exchanged, the devices create *logical channels* (each controlled by a single primary device). Devices use a Data Link layer protocol to tell others about themselves and to detect the presence of devices offering a service, to check data flow, and to act as a multiplexer. The standard also defines the packet structure.

The range of IrDA communications is between 10 centimeters and 1 meter (39 inches) although you can increase this range considerably when you increase the power of the device. The data transfer rate is from 9600 bps to 4 Mbps although originally the standard was 115 Kbps. The communication is always half-duplex. IrDA is well-suited to devices such as cell phones, mice, and keyboards because these devices consume a low amount of power.

When you were a kid at camp, after lights out, you may have used Morse code (does anyone still know Morse code?) to send messages to a buddy in the next tent. Well, to some extent, infrared works the same. IrDA devices communicate by using timed pulses of infrared light. The device employs light-emitting diodes (LEDs), which means you need line-of-sight to work. (If you want to see where infrared light fits in the spectrum, see Appendix C.) By turning light on and off at modulated times, you can send data. It uses the non-visible infrared light spectrum as its communications medium. For two IrDA devices to communicate using via infrared, you must point the infrared transceivers at each other, usually spaced no more than one meter apart.

Bluetooth, on the other hand, uses radio waves, which doesn't require a visual line-of-sight. Try this to see what we mean: Hold one hand up and shine a flashlight at it. Can you see the light on the other side of your hand? No, your hand absorbed the light. Now, hold up your hand and then hold up a radio behind it. Can you hear the radio program behind your hand? Of course, because your hand did not absorb all the radio waves. Also, the radio waves diffract around your hand. (See Appendix C for an explanation of diffraction and the nature of radio frequency.)

IrDA also doesn't work well in a well-lighted environment. Your office lights flicker at 60 Hz because of the alternating current used to power them, but you do not perceive it because your brain compensates. If the light is too bright, the flickering can interfere with your infrared signal. This is acceptable for remote controls but not for transmitting data. At least with Bluetooth and 802.11b and g, we just have to worry about interference from cordless phones, microwave ovens, and baby monitors. With infrared, you have to worry about lights. Sheesh.

Generally, you don't need to install any hardware to use infrared wireless ad hoc networking. Look at your cell phone or laptop, and you should see some red plastic. On the laptop, you may find it on the front, the back, or either side. One thing we know is that you won't find it on the bottom. Look at the top of your cell phone; you should see some red plastic there. This red plastic is your transceiver. The infrared *transceiver* is the small red window on your portable computer, printer, camera, dongle, or other device. If you find that you don't have a transceiver and want one, you will need to install one.

Installing infrared devices

Most internal IrDA devices are installed by Windows setup or when you start Windows after adding one of these devices. However, when you attach a serial IrDA transceiver to a serial (COM) port, you do need to install it in Windows. This section also describes how to install an internal IrDA device that is not detected by Windows, and how to reconfigure a serial port as an infrared port.

Installing an IrDA device connected to a serial port

If you have a desktop computer or a laptop computer without a built-in IrDA device, you can connect a serial IrDA transceiver to a serial (that is, COM) port. To install, attach the IrDA transceiver to the serial port, note the COM port you used, and then follow the following steps to add the new infrared device:

1. **From the Start menu, choose Settings⇨Control Panel and then open Add/Remove Hardware.**

2. **On the Welcome to the Add/Remove Hardware Wizard page, click Next.**

3. **On the Choose a Hardware Task page, select Add/Troubleshoot a device and then click Next.**

   You may have to wait while the wizard searches for your Plug and Play hardware.

4. **In Devices, click Add a new device, and then click Next.**

5. **On the Find New Hardware page, select No, I want to select the hardware from a list, and then click Next.**

6. **In Hardware types, click Infrared devices and then click Next.**

7. **In Manufacturers, click the manufacturer, and in Infrared Device, click the infrared device.**

8. **If you have an installation disk for the infrared device, click Have Disk.**

9. **Click Next, and then follow any additional instructions to install the device.**

After you add your infrared device, you may have to restart your computer before you can select the infrared port and device you just added.

Installing an undetected internal IrDA device

If you add an internal IrDA device to a computer with Windows plug-and-play (PnP), your system normally detects and installs the device the next time you start the computer. If this does not occur, you may have to install the device manually. To do this, refer to the preceding procedure.

This procedure installs an infrared device when your system does not support a separate infrared port. Some desktop computers allow you to reconfigure a serial port as an infrared port, which normally enables the computer to use Plug and Play to install the device.

Reconfiguring a serial port as infrared

On some desktop computers, you can reconfigure a serial port as an infrared port. You can use this to specify one of the COM ports as an infrared port.

Use this procedure only for an internal IrDA device. Do not perform this procedure to connect a serial IrDA transceiver to a serial port because the procedure disables the serial port.

After you perform the procedure, Plug and Play should detect the infrared device when you run the Add/Remove Hardware Wizard or after you restart the computer. For additional details, you should refer to your manufacturer's documentation provided with the computer or the infrared device.

Using IrDA to transfer data

Using IrDA is almost as easy as installing it. In Windows 2000, you choose Start⇨Settings⇨Control Panel. Double-click the Wireless Link icon. The Wireless Link dialog box appears (see Figure 4-1).

From the File Transfer tab, you see the default options. Basically, your system is wide open. At least when you select the first option, you'll know when people are connecting to you. If you decide to allow others to beam files to you, you should direct them to a secure location on your system.

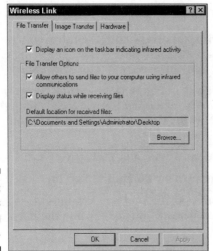

Figure 4-1:
Wireless
Link dialog
box.

Click the Hardware tab. You see a list of infrared devices on your system. The default is highlighted, but select the one you want to look at. Click the Properties button. The Infrared Port Properties dialog box appears (see Figure 4-2).

The General tab should be the active tab. If not, select it. At the bottom of the dialog box, you see the Device Usage drop-down list box. The system should have the device enabled by default, but you can enable or disable it here.

Figure 4-2:
Infrared Port
Properties
dialog box.

To establish an infrared link and make a network connection:

1. **Reposition your infrared transceivers until the infrared icon appears on your taskbar. Make sure that you have visual line-of-sight between the two devices and that the devices are in close proximity.**

2. **Choose Start⇨Settings⇨Control Panel. Double-click Network and Dial-up Connections. You also can open Network and Dial-up Connections by double-clicking Network and Dial-up Connections in My Computer.**

3. **Double-click Make New Connection, and then click Next.**

4. **Click Connect Directly to Another Computer, and then click Next.**

If Connect My Computer Directly to Another Computer does not appear in the Network Connection Wizard, you need to add the infrared device to the computer.

5. **To indicate whether this computer is sending or receiving files, do one of the following:**

   • *To initiate a connection,* click Guest.

   • *To receive a connection,* click Host.

6. **Click Next.**

7. **Under Select a Device, click Infrared Port, and then click Next.**

8. **To make the device available to all profiles, click For All Users, and then click Next. Or, to make the device available to just the current profile, click For Myself, and then click Next.**

9. **If this computer is a host, select the Users Allowed To Use This Connection, and then click Next.**

10. **Enter a name for the connection, and then click Finish.**

To examine or change properties for this connection, right-click its icon in Network and Dial-up Connections.

Securing IrDA

The IrDA standard does not specify security measures for data transfer. Because you require line-of-sight for data transfer, a low level security is provided. Don't point that thing unless you intend to use it! In that regard, infrared is more secure than Bluetooth and 802.11 technologies that are radio broadcasts.

For the most part, handheld devices currently have coarse-grained support for IrDA security. Basically, it is either on or off. Alternatively, you can enable or disable the port. Remember from earlier in this chapter that the default for infrared support is enabled.

IrDA depends on application level security measures for tight security. Therefore, your application developers need to implement authentication, encryption, or other security measures when needed.

There was a Windows 2000 denial of service attack based on buffer flow using the IrDA port, but you are fully patched, so no problem. Right?

There is even an infrared crack available on the Internet. Beamcrack is a simple application that will set or reset the bit in each application's database header that tells the launcher that it is or isn't *beamable,* thus bypassing the Palm Pilot's copy-protection. You can download Beamcrack from `www.10pht.com/~kingpin/beamcrack.zip`.

IrDA fills a networking niche up to one meter. WLANs are great for 10–100 meters. Bluetooth steps into the breach to fill the gap between 1 and 10 meters. Its ideal for ad hoc file sharing in a boardroom or anywhere you have not set up a wired or wireless network.

Understanding Bluetooth

Essentially, Bluetooth (`www.bluetooth.com`) is an ad hoc networking technology. *Ad hoc networks* have no fixed infrastructure, such as base stations or access points. In ad hoc networks, devices maintain random network configurations formed impromptu. Devices within the ad hoc network control the network configuration and maintain and share resources. Ad hoc networks allow devices to access wireless applications, such as address book synchronization and file sharing applications, within a Wireless Personal Area Network (WPAN). When combined with other technologies, you can expand these networks to include intranet and Internet access. Bluetooth devices that themselves do not have access to network resources but are connected in a Bluetooth network with an 802.11 capable device can connect wirelessly to your corporate network as well as to the Internet.

Ad hoc networks today are based primarily on Bluetooth technology. Bluetooth is an open standard for short-range digital radio. Its strong points are that it is a low-cost, low-power, and low-profile technology that provides a mechanism for creating small wireless networks on an ad hoc basis. Bluetooth is considered a wireless PAN technology that offers fast and reliable transmission for both voice and data. Bluetooth devices will eliminate the need for cables and can provide a bridge to existing networks.

Bluetooth is designed to operate in the unlicensed ISM (industrial, scientific, medical) band that is generally available in most parts of the world. This is the spectrum from 2.4 to 2.4835 GHz. 802.11b and g share this bandwidth. Because numerous other technologies also operate in this band, Bluetooth uses the aggressive full-duplex Frequency Hopping Spread Spectrum (FHSS) with Gaussian Frequency Shift Keying (GFSK) modulation in the range to solve interference problems. It hops 1,600 times per second and uses 79 different radio channels. The communicating devices will make use of one channel for 625 microseconds and then hop in a pseudo-random order to another channel for another 625 microsecond transmission; repeating this process continuously.

Bluetooth networks can support either one asynchronous data channel with up to three simultaneous synchronous speech channels or one channel that transfers asynchronous data and synchronous speech simultaneously.

There are two modes for the radio: asymmetric and symmetric. For asymmetric, the theoretical maximum data rate is a relatively low 1 Mbps with a throughput of 721 Kbps in one direction and 57.6 Kbps in the other. For symmetric, you get 432.6 Kbps in both directions. The difference between the throughput and data rate is due to the communication overhead. Regardless of the mode, the data rates and throughput are comparable with a typical Internet connection. The second generation of Bluetooth technology is expected to provide a maximum bandwidth of 2 Mbps. The data rates seem low especially when you compare them with 802.11 wireless LANs, but the data rate is still three to eight times the average data rate of parallel and serial ports, respectively.

Many books will go on and on about how Bluetooth will interfere with 802.11b and g because they both use 2.4 GHz ISM band. (In fact, we do this later on in the book.) Truth be told, it's not that bad. You can use Bluetooth alongside 802.11b or g with minimal interference. Devices such as Apple's PowerBook include both technologies onboard, so they must have worked out a solution to allow both to work. Right now, the workstation used to write this chapter has both Bluetooth and 802.11g clients. The 802.11 client utility shows the signal strength as 46 dB — an excellent signal. More important, the data rate is still the maximum, and there are very few packets retried. Each and every one a good sign. All things considered, this is a very strong signal with no significant frame loss. Shutting down the Bluetooth adapter provides little appreciable increase in signal strength or has any effect on frame loss. So, use both technologies because they are really complementary and solve very different problems. Though we see few co-existence problems, manufacturers of both Bluetooth and 802.11 equipment recommend that you not put transceivers within three feet of each other. Some manufacturers are starting to use adaptive frequency hopping spread (AFHSS) spectrum to help with co-existence. AFHSS will change the hopping sequence when encountering interference in any part of the band. Intel purchased Mobilian (www.mobilian.com), a manufacturer that had a chipset that handled 802.11 and Bluetooth simultaneously.

The operating range is about 10 meters (or 30 feet), but you can extend it to 100 meters (using more power). Up to 10 meters is considered your personal operating space for networking, so these devices work in your personal operating space.

Bluetooth provides three classes of power management:

- **Class 1 devices:** These are the highest power devices, operate at 100 milliwatt (mW), and have an operating range of up to 100 meters (m).

- **Class 2 devices:** These operate at 2.5 mW and have an operating range of up to 10 m.

- **Class 3 devices:** These are the lowest power devices, operate at 1 mW, and have an operating range of from one-tenth meter to 10 meters. This range is good enough for applications such as cable replacement (for example, mouse or keyboard), file synchronization, or business card exchange. Additionally, as with the data rates, you will see even greater distances in the future (again, more power).

You can use Bluetooth to connect almost any device to any other device. An example is the connection between a PDA and a mobile phone. The goal of Bluetooth is to connect different devices — for example, PDAs, cell phones, printers, and faxes — together wirelessly in close proximity such as your office, car, or home. Bluetooth was originally designed primarily as a cable replacement protocol for wireless communications. Among the assortment of devices you will see are cellular phones, PDAs, notebook computers, laptop computers, modems, cordless phones, pagers, cameras, PC cards, fax machines, and printers.

Bluetooth is now standardized within the IEEE 802.15 Personal Area Network (PAN) Working Group that formed in early 1999. See Appendix B for information on standards. Note that not all Bluetooth devices are 802.15-compliant. However, you should find it easy to upgrade Bluetooth-compliant devices to make them 802.15.1-compliant.

Bluetooth-enabled devices will automatically locate each other, but making connections with other devices and forming networks may require user action. Sometimes they connect automatically, which is a feature called *unconscious connectivity.*

Like with all ad hoc networks, Bluetooth devices establish connections on a temporary and random basis. A distinguishing feature of Bluetooth networks is the master-slave relationship maintained between the network devices. You can network up to eight Bluetooth devices together in a master-slave relationship, called a *piconet.* In a piconet, one device becomes the designated master for the network with up to seven slaves directly connected. The master device controls and sets up the network, which includes defining the network's hopping scheme. The master may have a total of 256 connections, but only seven can be active at any time. A master can suspend its connection to a slave by parking it and taking another slave. Devices in a Bluetooth piconet operate on the same channel and follow the same frequency hopping sequence.

Although only one device may perform as the master for each network, a slave in one network can act as the master for other networks, thus creating a chain of networks. And, a device can act as a slave in two piconets. By linking a series of piconets, you can create *scatternets,* which allow the internetworking of several devices over an extended distance. This relationship also allows for a dynamic topology that may change during any given session: As a device moves toward or away from the master device in the network, the topology and therefore the relationships of the devices in the immediate network change. Figure 4-3 shows the relationship of piconets and scatternets.

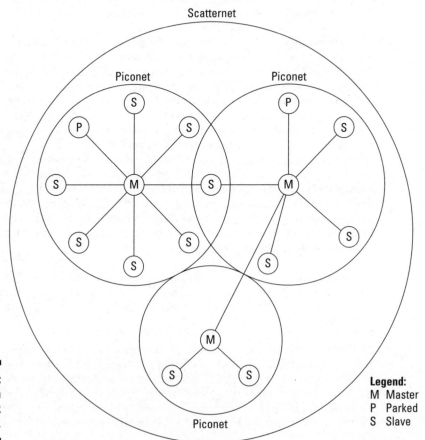

Figure 4-3:
Bluetooth
network
topology.

Legend:
M Master
P Parked
S Slave

Unlike a WLAN that comprises both a wireless station and an access point, with Bluetooth, there are only wireless stations or clients. A Bluetooth client is simply a device with a Bluetooth radio and Bluetooth software module with the Bluetooth protocol stack and interfaces.

Adding Bluetooth capabilities

Bluetooth offers five primary benefits to users. This ad hoc method of unfettered communication makes Bluetooth very attractive today and can result in increased efficiency and reduced costs. The efficiencies and cost savings are attractive for the home user and the enterprise business user alike. So, you may want to install Bluetooth to share your files and printers or to allow someone the use of her keyboard 10 feet from the desktop.

Using Bluetooth with Linux

Making Linux work with Bluetooth is not as straightforward as making it work with Windows or Mac OS. First, you will find three major and different Bluetooth stacks for Linux. Your first task is to ensure that you have a supported product. The most popular stack is BlueZ. You can find information about supported products at www.holtmann.org/linux/bluetooth/devices.html. You can find supported product information for Affix software at bthow.sourceforge.net/html-nochunks/howto.html. And finally, you can find supported products for OpenBT at sourceforge.net/projects/openbt. Affix and BlueZ are available under GNU Public License (GPL).

After you determine that you have drivers for your device, you will need to determine that your distribution of Linux supports Bluetooth. You can test your kernel by trying modprobe rfcomm as root. A positive response is good news. If you get bad news, try rebuilding your kernel to version 2.4.21 or higher and select all the options for Bluetooth support. To be safe, read the man page. If you are using Red Hat 9.0 or higher, you will find good Bluetooth support, including some BlueZ utilities.

Installing and using Bluetooth with Windows

There are many Bluetooth vendors, so there are many different ways to install Bluetooth. Microsoft provides a software package that provides hardware makers with a standard interface. Microsoft provides support for Bluetooth starting with Microsoft Windows XP SP1 and Windows CE. However, most hardware makers have chosen not to use Microsoft's software. A good example is BlueGear, which sells a wireless home network USB twin pack. You get two Bluetooth 1.1- and USB 1.1-compliant devices — little blue devices with 1.5 inch vertical antennas that plug into the USB port. You can use a BlueGear network to share an Internet connection, to share MP3 or other files, to print documents, and to play MUDs. BlueGear works with Windows 2000 and XP (and Me and 98 SE, for that matter).

To install BlueGear is simple. Follow these steps:

1. **Insert the CD and install the BlueGear software. If you turned off Autorun on your CD/DVD drive, choose Start⇨Run and browse the CD looking for the Setup program.**

2. **Follow the setup instructions. You will need to restart your system.**

3. **Plug the BlueGear into the USB port.**

   Your system will detect the hardware and install the drivers for you.

4. **Start the BlueGear applet. You open the applet by double- or right-clicking the blue starfish in the system tray.**

   You will see the icon in Figure 4-4.

BlueGear icon

Figure 4-4:
BlueGear
icon.

5. **On startup, you will need to enter a passkey and confirm it. Click Select Join the BlueNetwork from the menu.**

6. **Click the Search button.**

   You will see a computer and magnifying glass icon beside the Search BlueNetwork(s) title. When it stops, you will see a list of Bluetooth devices. Figure 4-5 shows the BlueNetwork dialog box with a found device.

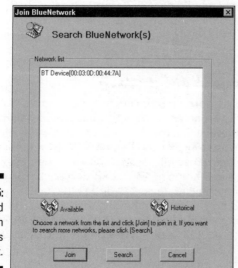

Figure 4-5:
Found
Bluetooth
Devices
dialog box.

7. **Highlight the device where you want to connect, and then click the Join button.**

8. **When you are connected, the blue starfish will rotate.**

To setup security, right-click the BlueGear icon (the blue starfish), select Options from the menu, and do the following:

1. **Select the Use Fixed Passkey box.**

2. **Enter a passkey in the Passkey box. Confirm the passkey in the Confirm box.**

3. **Click the Advanced button.**

    You will see the dialog shown in Figure 4-6.

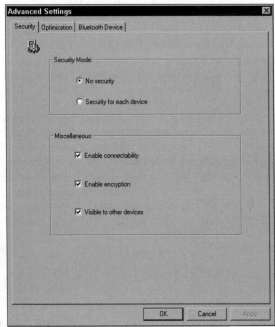

Figure 4-6:
BlueGear
Advanced
Settings
dialog box.

4. **Click the Security for each device radio button and click Apply.**

5. **Click OK.**

6. **Click Close.**

You now have a passkey for your device that you will need to share with all the Bluetooth device owners when you want them to connect. You can also create pairings under Options. These last two options aren't exactly self-evident, so in the next section, we look at Bluetooth security features.

Securing Bluetooth

Like any network, Bluetooth-based networks are susceptible to attacks. The types and volume of attacks should increase as more and more people deploy the technology. Currently, most of the attacks involve cell phones because that is where Bluetooth is most widely used. As this changes, so will the threat vectors and targets.

Toothing, bluesnarfing, Red Fang, and other attacks

Early versions of Bluetooth had security issues, but it looks like they are still coming. Bluetooth version 1.2 has a problem with how it deals with the personal identification number (PIN) that's used to protect data. You can break the identifier by using specialized hardware to capture certain data transferred between Bluetooth-enabled devices when they first contact each other. The hardware for cracking Bluetooth signals would cost you more than $15,000. (Dollar amounts are US.) However, you could turn some programmable wireless cards costing less than $1,000 into Bluetooth-eavesdropping equipment. The cracker has to eavesdrop on the initial negotiation between two Bluetooth devices, called *bonding*. After the information is collected, an eavesdropper can listen to cell phone calls, grab personal information as you synchronize with your computer, or counterfeit signals from one device to another.

The would-be eavesdropper would have to collect sufficient key data during the bonding process to have enough information to crack secret PIN codes. How much data depends on the number of digits you use for your PIN. An attacker can break a 6-digit PIN in a little over 10 seconds, whereas a 16-digit PIN would take more than 2,739 years or over a million days to crack. Alas, many Bluetooth-enabled headsets use 4-digit PINs that an attacker can break in less than a second. Your organization can defend its devices by selecting PIN passwords with a 10-digit password that would take literally weeks to crack.

 If you use short PINs, you are exposing data on the device. In addition, your Bluetooth users should avoid initially connecting their devices in a public place to limit the information a potential attacker could collect. If you are truly paranoid, then just keep moving!

On the other hand, someone doing surveillance of your Bluetooth-enabled devices is harder to foil. Using inexpensive electronics, anyone could create a Bluetooth device that could detect your device as far as a kilometer away, allowing them to track you via your cell phones. Alas, there is nothing you can do to prevent the tracking, other than to disable Bluetooth.

Now, our discussion gets colorful. We hope you are not easily offended. *Red Fang* exposes the location of hidden Bluetooth devices, and *bluestumbling* (also known as *bluesnarfing*) allows an attacker to grab information from certain makes of phones (some, but not all, Nokia, Ericsson, and Sony Ericsson handsets) that have poorly implemented security. Red Fang (www.atstake. com/research/tools/info_gathering) is an application that helps you to find non-discoverable Bluetooth devices by brute-force. We mention *war driving* numerous times in this book. Well, someone has coined this technique as *war nibbling,* which is the process of mapping Bluetooth devices within an organization. The Shmoo group also provides Bluesniff at `www.shmoo.com/projects.html` for device discovery. Perhaps, someone will develop Sweettooth in the future as a honey pot to attract all those war nibblers.

Bluesnarfing or bluestumbling allows you to bypass the pairing process to connect to a Bluetooth-enabled phone and essentially break into the device to steal or manipulate data. In short, somebody with the right program on their laptop within 10 meters can remotely discover your device, create a connection with no confirmation or code-input needed, and download your contacts and calendar to their computer. But it's not so easy. The bluesnarfer must stay within 10 meters for 2 or 3 minutes. Imagine trying to keep someone in range for that long. Just look for someone running after you as you head for your commuter train or head for the washroom.

Your organization must develop a Bluetooth policy. The policy most likely will depend on the device. For phones, you may want to set your Bluetooth to *undiscoverable.* For other devices, you may want to turn Bluetooth off completely unless it is absolutely needed. Whatever you choose, develop a policy and communicate it to all staff.

Protecting Bluetooth networks

Briefly, the three basic security services defined by the Bluetooth specifications are authentication, confidentiality, and authorization. As with the 802.11 standard, Bluetooth does not address other security services such as audit and non-repudiation. If you require these other security services, then you must provide them through other means. We describe here the three security services offered by Bluetooth and details about the modes of security.

Why *Blue*tooth?

Bluetooth, why Bluetooth? Why not gold tooth? Or, why not silver amalgam tooth? Ericsson Mobile Communication, the original architect for Bluetooth, named the technology after the tenth century (940-986 AD) Danish king Harald "Bluetooth" Blatånd II, a renowned communicator. He also was known as a unifying force in Europe in that century. Now, Danish isn't our specialty, but Blaatand is Bluetooth in English. Perhaps, because Bluetooth was the first in a line of Danish royalty, a unifier and a good communicator, they envisioned this communications technology as the first of a long line of technology that will unify devices like your wireless mouse and your desktop computer.

Also worthwhile to note is that Bluetooth is a frequency-hopping technology with 1,600 hops/second combined with radio link power control to limit transmit range. These features provide Bluetooth with some additional, but insufficient, protection from eavesdropping and malicious access. The frequency-hopping scheme, primarily a technique to avoid interference, makes it slightly more difficult for an adversary to locate a Bluetooth transmission. Using the power control feature appropriately forces any potential adversary to get up-close and personal.

Security features of Bluetooth per the specifications

Bluetooth provides three modes of security (none, service level, and link-level), two levels of device trust, and three levels of service security, stream encryption for confidentiality, and challenge-response for authentication.

To start, Bluetooth has three different modes of security. A Bluetooth device can operate in only one mode at a time. The three modes are the following:

- **Mode 1, Non-secure mode (no security):** In this mode, a device will not initiate any security procedures. In this non-secure mode, authentication and encryption are completely bypassed. In effect, the Bluetooth device in Mode 1 is in "promiscuous" mode that allows other Bluetooth devices to connect to it. This mode is provided for applications where you don't require rigorous security, such as exchanging business cards.

- **Mode 2, Service-level enforced security mode (L2CAP):** In this mode, the service-level security mode, security procedures are initiated after channel establishment at the Logical Link Control and Adaptation Protocol (L2CAP) level. L2CAP resides in the Data Link layer and provides connection-oriented and connectionless data services to upper layers. For this security mode, a security manager (as specified in the Bluetooth architecture) controls access to services and to devices. The centralized security manager maintains policies for access control and interfaces with other protocols and device users. You can define various security policies and "trust" levels to restrict access for applications with different security requirements operating. Therefore, you can grant access to some services without providing access to other services.

- **Mode 3, Link-level enforced security mode (PIN authentication/MAC address security/encryption):** In this mode, the link-level security mode, a Bluetooth device initiates security procedures before establishing the channel. This mode supports one-way or mutual authentication and encryption. These features are based on a secret link key shared by a pair of devices. To generate this key, the devices use a pairing procedure when they communicate for the first time.

Bluetooth bonding

The link key is generated during an initialization phase, while two Bluetooth devices that are communicating are *associated* (or bonded). Per the Bluetooth specification, two associated devices simultaneously derive link keys during the initialization phase when a user enters an identical PIN into both devices. After initialization is complete, devices automatically and transparently authenticate and perform encryption of the link. It is possible to create a link key by using higher layer key exchange methods and then import the link key into the Bluetooth modules. The PIN code you use in Bluetooth devices is between 1 and 16 bytes. The typical 4-digit PIN may be sufficient for some applications; however, you may need longer codes for others.

Authentication

The Bluetooth authentication procedure is in the form of a challenge-response scheme. Two devices interacting in an authentication procedure are referred to as the *claimant* and the *verifier*. The verifier is the Bluetooth device validating the identity of another device. The claimant is the device attempting to prove its identity. The challenge-response protocol validates devices by verifying the knowledge of a secret key — a Bluetooth link key.

The steps in the authentication process are the following:

1. The claimant transmits its 48-bit cleartext address to the verifier.

2. The verifier transmits a 128-bit random challenge to the claimant.

3. The verifier uses the algorithm to compute an authentication response using the address, link key, and random challenge as inputs. The claimant performs the same computation.

4. The claimant returns the computed 32-bit response to the verifier.

5. The verifier compares the response from the claimant with the response that it computes.

6. If the two 32-bit response values are equal, the verifier continues connection establishment.

If authentication fails, a Bluetooth device waits a set amount of time before making a new attempt. This time interval increases exponentially to prevent an adversary from repeated attempts to gain access by defeating the authentication scheme through trial-and-error with different keys. However, it is important to note that this suspend technique does not provide security against sophisticated adversaries performing offline attacks to exhaustively search PINs.

Again, the Bluetooth standard allows both one-way and mutual authentication. The authentication function uses the SAFER+ algorithm for the validation.

Avoiding (or not avoiding) "hooking up" using your cell

A new high-tech trend in Britain is called *toothing,* but it has absolutely nothing to do with dentistry. Toothing allows people to "hook up" using their cell. It's called *toothing* because Bluetooth wireless technology makes it all possible. You can anonymously send your best pickup lines to other Bluetoothers within 10 meters. You can find the "Beginner's Guide To Toothing" on a blog at (toothing.blogspot.com/2004_03_01_toothing_archive.html) dedicated to the pursuit. Jon, also know as *Toothy Toothing* (and the guide's author) explained that he conceived the idea after he was "bluejacked" by an unknown young lady while commuting to work in London.

Toothing sounds all right, but bluejacking sounds painful. Toothing is facilitated by jacking. Now, that sounds rude. *Bluejacking* is a craze where people send anonymous messages to other people using Bluetooth equipment. To bluejack, you

1. **Find a Bluetooth-enabled device such as a mobile phone, PDA, or laptop. Generally, this means a Bluetooth-enabled phone.**

2. **Create a new phone book contact and the message you want to send to someone in the Name field. Put a three or four word message in the display area reserved for the name of the initiating device.**

3. **Find somewhere where there are likely to be other Bluetooth users.**

4. **Select the contact you made earlier, and choose Send via Bluetooth.**

5. **Your phone will search for available Bluetooth devices within 10 meters of you. It will either list available devices or say none were found. If the latter, find a better or busier spot.**

6. **From the names of devices in range, select one to receive your phone book contact.**

7. **If all goes well, your phone will send your contact to the selected device.**

8. **Try to casually look around you and see whether you spot anybody looking at their Bluetooth-enabled phone and perhaps reading your message.**

9. **Well, that's it. Hope your message was urbane and sophisticated, or at least humorous. Guess this takes us back to toothing! When participating in toothing, you usually enter Toothing in the Name field.**

Bluetooth technology is an enabler that allows people to swap data between mobile phones, PDAs, notebook computers and other devices within a few meters of each other. That's point. So don't be surprised when it happens to you!

If you don't want to be bluejacked or toothed in public places, you should either switch your phone to the *non-discoverable* or hidden mode (making it invisible to others) or turn off Bluetooth completely. You should also check that your Bluetooth *pairings* (approved connections with trusted partners) are correct.

The Bluetooth address is a public parameter that is unique to each device. This address can be obtained through a device inquiry process. The private key, or link key, is a secret entity. The link key is derived during initialization, is never disclosed outside the Bluetooth device, and is never transmitted over the air.

The random challenge, obviously a public parameter, is designed to be different on every transaction. The random number is derived from a pseudo-random generator (PRNG) within the Bluetooth device.

The cryptographic response is public as well. With knowledge of the challenge and response parameters, it should be impossible to predict the next challenge or derive the link key.

Confidentiality

In addition to the authentication scheme, Bluetooth provides encryption to thwart eavesdropping attempts and protect the data exchanged between two Bluetooth devices.

The Bluetooth encryption procedure is based on a stream cipher. A key stream output is exclusive-OR-ed with the payload bits and sent to the receiving device. This key stream is produced using a cryptographic algorithm based on linear feedback shift registers (LFSR). The encrypt function takes as inputs the master identity, the random number, a slot number, and an encryption key, which initialize the LFSRs before the transmission of each packet, when encryption is enabled. Because the slot number used in the stream cipher changes with each frame, the ciphering engine is also reinitialized with each frame although the other variables remain static.

An internal key generator produces the encryption key provided to the encryption algorithm. This key generator produces stream cipher keys based on the link key, random number, and the ACO value. The *ACO value,* a 96-bit authenticated cipher offset, is another output produced during the authentication procedure. As mentioned previously, the link key is the 128-bit secret key that is held in the Bluetooth devices and is not accessible to the user. Moreover, this critical security element is never transmitted outside the Bluetooth device.

The encryption key is generated from the current link key. The key size may vary from 8 bits to 128 bits and is negotiated. The negotiation process occurs between master devices and slave devices. During negotiation, a master device makes a key size suggestion for the slave. In every application, a "minimum acceptable" key size parameter can be set to prevent a malicious user from driving the key size down to the minimum of 8 bits, making the link totally insecure.

The Bluetooth specification also allows three different encryption modes to support the confidentiality service:

- **Encryption Mode 1:** No encryption for any traffic.

- **Encryption Mode 2:** Broadcast traffic goes unprotected (not encrypted), but individually addressed traffic is encrypted according to the individual link keys.

- **Encryption Mode 3:** All traffic is encrypted according to the master link key.

Trust levels, service levels, and authorization

In addition to the three security modes, Bluetooth allows two levels of trust and three levels of service security. The two levels of trust are *trusted* and *untrusted.* Trusted devices are ones that have a fixed relationship and therefore have full access to all services. Untrusted devices do not maintain a permanent relationship; this results in a restricted service access.

For services, three levels of security have been defined. These levels are provided so that the requirements for authorization, authentication, and encryption can be set independently. The security levels are as follows:

- **Service Level 1:** Require authorization and authentication. Automatic access is granted only to trusted devices. Untrusted devices need manual authorization.

- **Service Level 2:** Require authentication only. Access to an application is allowed only after an authentication procedure. Authorization is not necessary.

- **Service Level 3:** Open to all devices. Authentication is not required, and access is granted automatically.

Associated with these levels are the following security controls to restrict access to services:

- **Authorization required.**

  This always includes authentication.

- **Authentication required.**

- **Encryption required.**

  Link must be encrypted before the application can be accessed.

The Bluetooth architecture allows for defining security policies that can set trust relationships in such a way that even trusted devices can get access only to specific services and not to others. It is important to understand that Bluetooth core protocols can authenticate only devices and not users. This is not to say that user-based access control is not possible. The Bluetooth security architecture (through the security manager) allows applications to enforce their own security policies. The Link layer, at which Bluetooth-specific security controls operate, is transparent to the security controls imposed by the Application layers.

Thus, it is possible to enforce user-based authentication and fine-grained access control within the Bluetooth security framework.

Combating Bluetooth security problems

Bluetooth security problems arise because of the PRNG, short PINs, negotiable encryption key lengths, reusable and disclosed unit key, shared master key, no user authentication, unlimited authentication attempts, weak stream algorithm, and the simple shared-key challenge-response. If you want to research some of these problems, check out www.niksula.cs.hut.fi/~jiitv/bluesec.html, grouper.ieee.org/groups/1451/5/Comparison%20of%20PHY/Bluetooth_ 24Security_Paper.pdf, and www.giac.org/practical/gsec/Nikhil_ Anand_GSEC.pdf.

The following are some Bluetooth security countermeasures to address these weaknesses:

- ✔ Make sure that you turn off all Bluetooth devices when you are not using them to minimize the exposure opportunity.

- ✔ Set Bluetooth devices to the lowest necessary and sufficient power level so that transmissions remain within your perimeter.

- ✔ Ensure that Bluetooth bonding or key exchange is secure from eavesdroppers.

- ✔ Choose random and strong PIN codes to thwart guessing.

- ✔ Choose long PIN codes (say the maximum of 16) — not 4 digits, like your bank card.

- ✔ No Bluetooth device should default to the zero PIN (that is, 0000).

- ✔ Configure Bluetooth devices to delete PINs after initialization to ensure that you must re-enter the PIN every time.

- ✔ Ensure PINs are not stored in memory after power removal.

- ✔ Use combination keys instead of unit keys.

- ✔ Use link encryption for all Bluetooth connections.

- ✔ Use Security Mode 2 in controlled environments only.

- ✔ Use device mutual authentication.

- ✔ Enable encryption for all broadcast transmissions.

- ✔ Use the longest encryption key sizes allowed.

- ✔ Establish a minimum key size for any key negotiation process.

IrDA and Bluetooth Comparison

If you examine the benefits of each technology, you can see that Bluetooth and IrDA are both critical to the marketplace. Each technology has advantages and drawbacks, and neither can meet all your needs. IrDA is still a very active technology, but Bluetooth has emerged as the dominant wireless networking technology for distances of less than 10 meters. Bluetooth as 802.15 will continue to grow.

Do not mistake these two technologies as networks; they are only means of connectivity. Both infrared and RF (radio frequency, used by Bluetooth) are needed to solve all wireless needs. Although IEEE developed 802.11 standards to get rid of all the CAT5 cable in your organization, IEEE developed IrDA and Bluetooth to replace all the other cables in your office. Now, all you need is to get rid of those nasty power cables and bricks!

Chapter 5

Moving On to a Wireless LAN: Your Wireless Access Point

*I*n this chapter, you install and set up the basic equipment for wireless networking: your wireless access point. Having a wireless session without an access point is limiting in scope because with this setup, you can participate only with your peers. You need to take care of some critical items during installation. We show you how to decide where to install the access point and configure the device to work on your network and change all the defaults so the bad guys cannot get in. First, though, you need to make sure that you have all the parts you need to get started.

Parts Is Parts — Do You Have Them All?

Okay, now for an easy task. Do you have all the parts? It's a crying shame to start work and then find out that your vendor missed a part, isn't it? You get so far along, and wham! Well, the easy solution is to verify what you have against the packing data that your vendor provides.

Start by ensuring that all items are in the box and that they appear undamaged. Does the box look as if it were damaged in any way prior to your opening it? This can hide internal damage done to the wireless access point that isn't obvious to you until you attempt a connection. If you are unsure and the container looks damaged, return it before you open it.

The small amount of time lost returning it can be more than worth it if the device is compromised in some way that might take far longer to troubleshoot.

A typical packing box contains the following items:

- ✓ The wireless access point
- ✓ A power adapter
- ✓ Three to six feet of RJ45 Category 5 (CAT5) cable
- ✓ An antenna or two depending upon the model
- ✓ The quick start guide
- ✓ An Easy Start CD with Installation Wizard and Manual

If you are using an antenna that you purchased separately for your access point, you also need a few feet of coax cable with connectors to attach it. Be sure that the cable is included with your antenna.

Connecting and Configuring Your Access Point

After you verify that you have all the components and assess them for damage, you can establish your wireless connection. You need to configure the wireless access point in order to use it on your network. Configuration involves a number of things, including the initial configuration that occurs when you power on the device; further configuration setting the IP address, netmask, and Domain Name System (DNS) servers to work on your organization's network; and perhaps configuring a firewall and other options, depending on the type of device and its supported options.

Note the difference between access points and routers. You might need one or the other or both for your network, depending on what you want to do. We discuss routers and access points in this chapter. If you're operating a small business, you may purchase a router rather than an access point as a gateway to the Internet.

After everything is out of the box, you need to connect all the parts and see whether you have wireless connectivity available. You typically use the quick install guide that is part of the package after you have connected all the parts.

We guide you through the steps for a SMC Wireless Router, model number SMC2804WBR. All wireless routers and access points have similar steps.

Connecting the access point

You need a power outlet nearby to connect the access point and a desktop with a working Ethernet card.

1. **Attach the two antennae to the back of the machine.**

   The antennae just screw on to the rather obvious-looking posts, one on each side of the router. If you purchased an external antenna, connect it now and place it in the location that provides the best signal.

   Place one antenna vertical and the other horizontal for best coverage.

2. **Attach the power supply to the back and plug in the power supply to a wall socket.**

   A green PWR light shows the device is receiving power.

   Consider labeling your wires while you connect them all, both electrical and Ethernet. Place an identifier on each end so that you easily recognize it. Labeling allows you to readily locate the one you need when you are troubleshooting connections.

3. **Attach a Category 5 Ethernet cable with RJ45 connectors to the back of the device in one of the connectors at the back typically numbered 1–4 or 1–8, depending on how many wired connectors the device allows.**

4. **Attach the other end of the Ethernet cable to your desktop's Ethernet card.**

   Access points like the CISCO 1100 and 1200 series also allow initial configuration by connecting through a serial port on the back of the device. You can connect a nine-pin, female DB-9 to RJ-45 serial cable to the RJ-45 serial port on the access point and to the COM port on your computer. Of course, this isn't required, and you can still use a LAN connection. You can see what we mean in Figure 5-1.

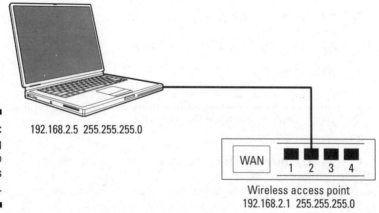

Figure 5-1: Connecting your PC to the access point.

192.168.2.5 255.255.255.0

WAN 1 2 3 4

Wireless access point
192.168.2.1 255.255.255.0

5. **Verify the connection is working by viewing the port lights on the front of the access point.**

   You see a light corresponding to the port number that you used in Step 3. This link light is green, signifying that the connection is established. The amber activity light indicates whether the connection is 10 Mbps when it is off, or 100 Mbps when it is lit.

Do not connect the wireless access point or router to your internal network. You need to change default passwords and put security in place before doing so. Also, ensure that your test desktop has no sensitive data on it because it is susceptible to attack until security is enhanced on the wireless network.

If you are using a router like the one we identify at the beginning of this procedure, you can also connect your cable or DSL modem connection to the WAN connector. This allows the wireless device to communicate with your Internet service provider's modem and allows wireless access to the Internet. Use the same Category 5 cable to perform this step. We recommend waiting, though, until you turn off all default access. You don't want to provide drive-by access to the Internet, do you?

Configure your browser

After you verify the connections, you need to configure your browser to disable any proxy settings that are in place to ensure that you reach your access point and not some other part of your network. To configure Internet Explorer 6.0, follow these steps:

1. **Choose Tools⇨Internet Options.**

2. **Select the Connections tab.**

3. **Click LAN Settings and clear all the check boxes in the window that displays.**

4. **Click OK twice to finish.**

To configure Netscape 4.7, follow these steps:

1. **Choose Edit⇨Preferences.**

   The Preferences dialog box appears.

2. **In the Category pane, click Advanced.**

3. **Select the Proxies option.**

4. **Select the Direct Connection to Internet option and then click OK.**

At this point, your wireless device is ready for connections using the system defaults.

Changing the default network settings

You may need to change the network settings on your connected desktop to access the device. Take note of the settings that you find in this step so you can change them back later to remain connected to your network if you don't use the wireless access point. In Windows XP, follow these steps:

1. **Choose Start➪Network Connections.**

   The Network Connections window opens.

2. **Select Local Area Network.**

   The Local Area Network dialog box opens.

3. **Select Properties and scroll down until you see TCP/IP. Double-click it.**

   The dialog box for setting your network connections opens.

   If the Obtain an IP Address Automatically option is selected, you have a useable address provided by the access point's DHCP service. Skip to Step 6 to test whether you obtained a valid address; otherwise, continue with setting up a static IP address.

4. **Select the Use the Following Address option.**

5. **Enter an IP address that matches your access points in the first three octets.**

   For example, the SMC device uses 192.168.2.1 as its address, so we enter **192.168.2** and any number between 2 and 254. We enter **192.168.2.5**, for example. Use your cursor to select the subnet mask. It should be 255.255.255.0. The gateway address can be set to your access point for now or left blank.

6. **Ignore the DNS fields, click OK, and then click Close to leave the configuration.**

   Make sure that your wireless access point and your network card are using the same subnet. Different vendors may specify different subnets. Not doing so causes you untold grief trying to connect.

7. **Test the connection by making sure that you can reach the device.**

   Open a DOS Command dialog box. At the prompt, type the following command, substituting the correct IP address for your specific device.

   ```
   ping 192.168.2.1
   ```

   If your network connections are accurate, you see a response indicated with four Reply From messages. If you do not get this message and see a Time Out message, you need to verify your network address again in Step 5. Figure 5-2 shows an example of a successful ping command.

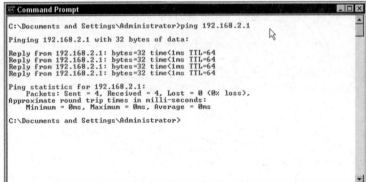

Figure 5-2:
Using
the ping
command
to verify
connectivity.

All your connections are in place and you verified that you are able to connect to the access point. We connect one workstation with this procedure. After this workstation is connected and running to your satisfaction, you can play with additional network connections.

Initial Setup and Testing

After you configure the device and it's available for access, you perform initial setup and testing. Here you implement the network settings needed to connect your users and set common parameters.

We guide you through some generic steps based on the SMC device mentioned earlier but with caveats for other device types.

Deciding on initial setup options

You need to connect to your access point and begin selecting options to set up the device. Here's how:

1. **Open your browser and type in the default IP address of the access point.**

   You'll find this address in the documentation. For the SMC device we use in the section "Connecting the access point," the address is 192.168.2.1. A main screen is shown with a login request, as shown in Figure 5-3.

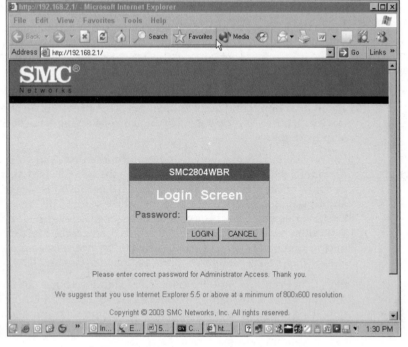

Figure 5-3:
Using the
main login
screen to
authenticate
to the
wireless
access point
or router.

2. **Some devices require you to enter the user account and password.**

   This account is usually called *Admin*. Consult your manual for your particular device. Enter the default password in the space provided along with a user account, if requested. Some devices use the word *default,* or have a blank password for the initial setup. Ugh! The SMC we are using has a blank default password! More on that later.

3. **After you enter the password and get connected, you see a menu offering choices.**

   Many of the security settings are discussed later in detail in Chapters 9 and 11. For now, continue with setting up the device.

4. **Click the Advanced Setup tab and select System. Under the System tab, select Time Zone.**

   Now set the time for the appropriate time zone you operate within to obtain the correct time on all your logs. If you have Internet access, leave the Enable Automatic Time Server maintenance field enabled so that you'll get the time from a Network Time Protocol (NTP) server. Set the optional server fields to those you already use for your internal servers or leave the defaults selected. Click Apply.

5. **Select Remote Management.**

   Leave the Password tab until later. Ensure that the Remote Management option is set to Disabled. We do not recommend allowing access to the device from remote networks. If this is set to Enable, users on the Internet can try and gain access by guessing your password. Figure 5-4 shows the main menu with the System tab selected.

6. **If you are connecting to an Internet service provider through the device, set the WAN options. If you are connecting an access point, skip to Step 8.**

   For broadband access, you need to inform the device of your service provider's details. There are various options here, from dynamically assigned addresses to using PPPoE and other methods. Select the one appropriate for your service provider and enter the details provided by them. For example, for broadband cable access in our area, we select the WAN entry for Dynamic Address IP address, select the More Configuration button, and enter the Host name provided by our service provider. Select Finish when you're ready to confirm all your settings and put them in place.

   PPPoE, or Point-to-Point Protocol over Ethernet, is used by some DSL providers to encapsulate PPP so it can be sent over an Ethernet connection. It's a method for specifying how a PC interacts with a broadband modem to get access to a high-speed network such as your ISP with little user interaction required.

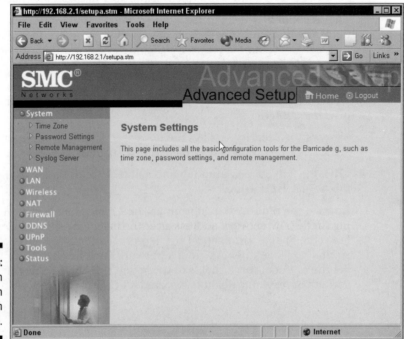

Figure 5-4:
The main menu with the System tab selected.

7. **For PPPoE users, enter the username and password provided by your service provider in the appropriate fields.**

   The Service Name is normally optional, but if it's required by your provider, enter it in the space provided. You can leave the Maximum Transmission Unit (MTU) at its default value unless your service provider has indicated otherwise. You also leave the Maximum Idle Time at its default unless you feel a need to modify it to a different time interval. Finally, enable the Auto-reconnect option to automatically re-establish a connection when you attempt to access the Internet. Clicking Finish sets the options you select.

   The other remaining service types include Fixed IP-xDSL and Point-to-Point Tunneling Protocol (PPTP), which is common in Europe. For Fixed IP-xDSL users, enter the assigned IP address, gateway, and DNS IP addresses, and finish by entering the subnet mask that is provided by your service provider. Europeans or others using PPTP need to enter the assigned IP address, subnet mask, default gateway address, username, password, and finally, their PPTP gateway address. You can also leave the Maximum Idle Time at its default unless you feel a need to modify it to a different time interval. Clicking Finish after either of these protocol settings are complete finishes the setup.

8. **Change your IP address for the device.**

   If you are using 192.168.2.0 as your internal network, then no worries, as they say in Australia. You can use the default settings. However, the majority of readers will likely have alternate addresses and therefore will need to follow these next few steps carefully.

9. **Select the LAN option.**

   In Chapter 2, you analyze your network. Now you need that information to determine the address you use in order for your users to connect to the wireless access point. Typical internal networks use a number of valid address ranges from 192.168.x.x to 10.x.x.x and many others. Enter the IP address you intend to use for the device in the field titled IP Address under the LAN IP heading.

   We recommend that you turn off DHCP because it is likely you already run a DHCP service in your network, and they could conflict. You disable it by selecting the Disabled check box. After you select the checkbox, the DHCP fields disappear. You see where to set these in Figure 5-5. Select Apply when ready. Note that you will need to restart your browser pointing to the new network address after the machine reboots.

10. **Remember to re-establish your workstation's network setting back to their original settings.**

    You did write down what they were prior to changing them didn't you? Because your access point now resides on the network address range, you need to reconnect using the same address range. The basic point is that your workstation needs to be on the same address range you use in Step 9.

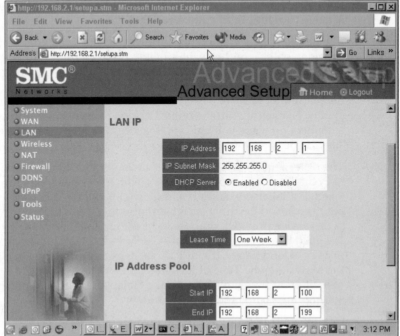

Figure 5-5:
Setting the
IP address
and
disabling
DHCP.

Performing the advanced setup functions

To finish getting connected, simply follow these steps:

1. **On some devices, you need to tell the device to perform as an access point.**

   You do this by setting a few parameters. You ensure the device acts as an access point by enabling it on the SMC. When it is reconnected, click Advanced Setup and then select the Wireless tab. You see two options indicating that the wireless function is Enabled or Disabled. Be sure that Enable is selected.

2. **In the main menu under Wireless, select Channel and SSID.**

   You see a menu like that shown in Figure 5-6.

3. **Find the field named SSID (shown on the SMC router as ESSID for Extended SSID).**

   This is the actual name that is broadcast to devices wanting to connect to the wireless network. Your wireless network card will require this name in order to connect. Leave it and the field called ESSID Broadcast set to its default for now, allowing this name to be sent across the wireless network.

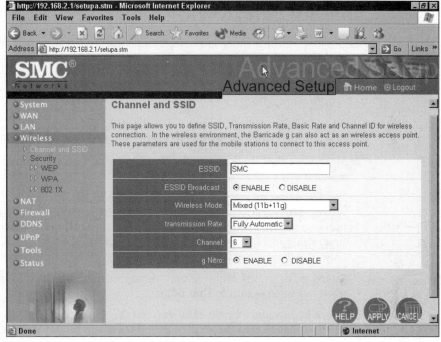

Figure 5-6:
Enabling
wireless
access and
setting the
SSID and
channel.

4. **Depending on your device, you may need to set the wireless protocol to be used, such as 802.11a, b, or g.**

   Our SMC allows both 802.11b and 802.11g traffic and is set for that as the default.

   Chapter 2 and Appendix B cover the difference between the protocols. Leaving the option set to allow both types allows you to use older and possibly cheaper wireless network cards. They do, however, operate at lower speeds. Finally, some devices like the SMC also allow for increased speeds specific to the vendor. Consult the manual that came with your access point for details on using vendor-specific speed options.

5. **If your device allows for alternate transmission rates, you can set these according to your needs.**

   For now, leave the default set. This is Fully Automatic on the SMC device.

6. **Set the channel you want devices to use for connections.**

   Your wireless network cards need to reflect the channel you want to use for connections and the SSID, or you will not get connected. The workstation or laptop's IP address must be the same subnet, as well. Leave this option set on channel 6, which is the default. Changing this option allows for less interference with other access points that may be within range of yours. Click Apply to set the changes in place.

Back up your work

Here are just a couple more important steps to take before going any further.

1. **Return to the Advanced menu. Skip down to the Tools tab, and select Configuration Tools.**

   The options in this tab allow you to back up your settings so you do not need to reapply them should a power failure or other event occur.

2. **Make sure that Backup Router Configuration is selected and select Next.**

   You are asked where you want to place the configuration file. Select an appropriate location and click Save. Your settings are saved to the location you specify. Should you need to reset the settings at some future time, simply navigate to this window, select the Restore check box, and enter the location of the backup.

   At any point in this process, you can select the Status option and verify what has been set up to that point.

3. **There is one more important step to take. Now that you can access device, ensure that the firmware it uses is the latest version.**

   On the SMC, the version is indicated on the Status page. It shows the Runtime Code Version. Visit your vendor's Web site and verify what the latest version is and download it if the one in your device is older.

4. **After the download is complete, from the main menu select Tools and then select Firmware Upgrade.**

   Select the Browse option and locate the new version you downloaded. Click Apply. You are prompted to confirm the upgrade.

5. **Click OK.**

   The device will update its firmware and reboot. Your device is enabled and ready to use. You verified each setting and ensured a degree of access as you tested the connection each time with your browser.

Turning Off the Defaults

Now you get serious about your wireless network. Setting up a wireless network is a good thing, but you need to be responsible and prevent unauthorized persons from using it. You can prevent someone from hacking in by changing the default settings.

Changing the password

One of the first items of business is changing the default password. Remember when you first set up your device, the password is blank. Blank. Empty. Void. Vacant. This is bad. You must create a new password that is difficult to guess and yet easy enough to use.

Follow these steps to set your new password.

1. **Log in to the device using the blank password or whatever the default is for your device.**

   You typically do this by pointing your browser to the correct IP address, where you will see the login page as we show in Figure 5-4.

2. **Navigate to the Advanced Setup tab and select it. Then select System.**

   This step will differ depending on the device you are using, so follow the instructions in your manual if you are unsure where to find the System options we discuss next. Next, under the System tab, locate the Password Settings tab and select that. Enter the old password (leave it blank on the SMC) and then enter a new, reasonably complex password, re-entering it to confirm in the final dialog box provided. Make sure that you remember the password! If you must write it down, which is not a good idea, keep the paper safely locked up. Do not key it into your workstation or laptop because too many people can access that and possibly see the password. Click Apply when you are done to complete the action. You need this password now in order to log in to the device.

   On the SMC device, you can modify the Idle Time Out parameter during this step. However, we recommend leaving the default unless you have an overriding need to change the amount of time before you need to login again due to inactivity. Do not change it to zero, however, because this means that your session remains open — possibly leaving you vulnerable to someone gaining access and changing settings should you leave your workstation unattended. After all, everyone visits the water cooler at some point. Play it safe and leave a time out value.

Changing the access point name

Consider changing your Service Set Identifier (SSID). This is the name that may get broadcast to identify your wireless access point. Using the default typically helps identify the type of device, possibly providing too much information to anyone who sees it. The SMC device used earlier in this chapter uses a SSID of SMC, and older models use the term *Default*. You change this to something that does not identify your organization, department, or anything useful. Why?

Again, it makes no sense broadcasting a name that gives others useful information. Change your SSID by following these steps:

1. **Log in to the device using your new password (see the preceding section).**

   You typically do this by pointing your browser to the correct IP address, where you will see the login page we show in Figure 5-3.

2. **Navigate to the Advanced Setup tab and select it.**

3. Select the Wireless option.

4. **Select Channel and SSID.**

   You see the default name provided by your vendor.

5. **Change the SSID field (shown on the SMC router as ESSID) to something innocuous, such as the names of stars.**

   This limits the degree of useful information passed to outsiders. You need to use this new name in all your wireless network cards in order to connect. Clicking Apply sets the field. Any wireless access card that might have been connected will now need to add that new name in order to reconnect, as described in Chapter 6.

Changing security options

We would be remiss if we did not mention that you need to set security so that only authorized users can connect to your new wireless network. There are a number of different methods for this. We show you the basic encryption method for now. You find out about the other methods in Chapter 11.

You turn on WEP, as you may have already noticed, through the same menu that we access throughout this chapter. You do not need to know a lot about encryption, but you do need to understand that the longer the secret is (the encryption key), the better the security. Therefore, in your wireless access point, use a 128-bit key rather than the older 64-bit key. When you do choose the higher bit key, however, your wireless network cards need to support the increased level of security. Several older and cheaper network cards support only 64-bit. If that is all you have, use them rather than not turning on WEP. After all, any security is better than no security.

Follow these steps to secure your wireless network.

1. **Log in to the device using the password you created.**

   You typically do this by pointing your browser to the correct IP address, where you will see the login page shown in Figure 5-3.

2. **Navigate to the Advanced Setup tab and select it.**

3. **Select the Wireless option.**

4. **Select Security.**

5. **Select WEP.**

   You see a number of options, as shown in Figure 5-7.

6. **You now see a field that presently sets the option to No Security.**

   It may appear as Disabled on some devices. You must enable this option in order for encryption to be used. In the SMC, select WEP from the drop-down dialog box.

7. **In the dialog box containing the WEP Mode, select 128-bit.**

8. **In the area indicated by the term Static WEP Key Setting, click the Clear icon to clear out any default settings.**

   You now need to generate a key that is between 10 and 64 hex characters long. Key it in the blank indicated by Key 1.

   What is a hex character you ask? Excellent question. So glad you asked. *Hex* is an arcane set of characters loved by old-style programmers. To us, it's simply the numbers 0–9 and the characters A–F. They correspond to the decimal numerals 0–15. You can find an interesting converter at `www.mikezilla.com/exp0012.html`.

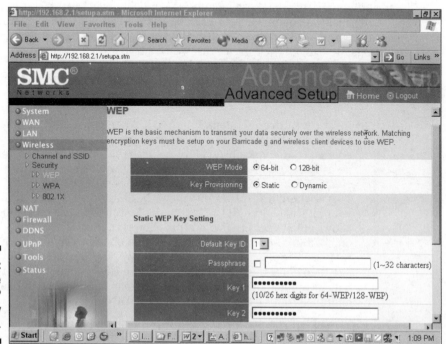

Figure 5-7:
Setting the
WEP
Security
option.

Unfortunately, before you start madly keying in random hex characters, you will need to repeat them once more here and then in every wireless network card. So write down the key you plan to use and then key it in. Keep that written copy safe because anyone seeing it has the keys to the wireless kingdom! Type in the key that you selected and then click Apply when you are done.

9. **You set the encryption key in the preceding step; now you need to tell the device to start using it. Select the Security option from the menu.**

   You see the device indicates there is no security. Change this to WEP Only by selecting it from the drop-down box. Click Apply when you are done to set the option. From this point on, you will need your wireless network cards to use this new level of security in order to connect.

10. **Log out of the wireless access point and then exit the browser.**

    In this section, we changed the default password and implemented security. Now you can connect the wireless access point to your network safely and allow users to begin connecting.

Understanding the other options

As you set up your SMC router, you may notice that it contains numerous additional settings. These include NAT and Firewall. Default settings for these are usually adequate for your use. We suggest working with the settings established previously in this chapter before adding the complexity of these settings. We explain their use in case you want to modify them later.

You use the settings in this section only if you have a router and attach to the Internet through the WAN port.

- **Network Address Translation (NAT):** When you connect to the Internet, you do not use your internal IP address range. One reason is because these addresses are typically non-Internet routable. In other words, they cannot be used across the Internet. Non-routable addresses include 10.0.0.0–10.255.255.255, 172.16.0.0–172.31.255.255, and 192.168.0.0–192.168.255.255. This is where NAT comes into play. The SMC router automatically changes your internal addresses to match the WAN IP address by default. This allows your access to the Internet. However, if you want to use more than one WAN address, select the NAT option on the menu and then select Address Mapping. Enter the additional WAN address in the Global IP field and then enter the range of user IP addresses that should use this alternate Internet address. You see that you can subdivide your users into multiple groups each with an assigned WAN address.

- **Virtual Server and Special Applications:** These tabs allow you to let users from the Internet access your internal machines. Be careful about using these options — they place your server on the Internet where it is susceptible to attack. A virtual server, for example, might be a Web server that you want the world to see. You place the IP address in the field provided and use port 80 for both the LAN and Public ports and then click Enable.

Special Applications include video conferencing or Internet telephony. You need to learn the ports that are used by these applications in order to complete their setup in this menu. We strongly recommend not using these applications unless absolutely necessary because they open up access to your internal network and can be used by hackers to attack your site. Besides, the common uses include games and ICQ, and you probably shouldn't be running such things in a business environment, anyway.

✔ **Firewall:** The firewall option is definitely one to use if you are connecting the WAN port. This helps protect your network from unauthorized users. It is enabled if the Enable box is marked. After you verify that it is, select Access Control if you wish to restrict which of your users can access the Internet. Should you choose to do this, you can apply the criteria for allowing access shown after you select Enable Filtering Function and select Add PC. Read the manual for explicit details on how to set this restriction.

Using the firewall allows you to implement Intrusion Detection. This uses rules to help prevent attacks against your network and is a good thing. The default settings will protect your site reasonably well. Further reading of the Help files can provide more details. Note that it allows you to automatically be notified by email if someone is attempting to hack your network.

✔ **MAC filtering:** This allows you to restrict wireless network clients based on the physical address in their network card. You find more details on doing this in Chapter 10.

✔ **URL blocking:** This page allows a fairly simplistic attempt at restricting the Web sites your users go to. You can type in up to 30 URLs or keywords that will be used to prevent access. You can also schedule the rules that you use in the Access Control section to apply at varying times.

✔ **Demilitarized Zone (DMZ):** This is a portion of your network that is open to the Internet. Adding a client to the DMZ will expose your local network to numerous security risks, so we do not recommend it.

Configuration and setup of a Cisco Aironet 1200

Previously in this chapter, we provide the steps needed on an SMC device. In this section, we provide a quick recap for those organizations using Cisco access points. The 1200 offers both 802.11b and 802.11a access (and g), and includes settings for each wireless mode. You find that these setting are similar in both the 1100 series and the 1200 series access points.

1. **Connect to your Aironet device using either a LAN connection in the same manner as the SMC or using the COM port.**

2. **Open your browser and connect using the default IP address of the device.**

   The device default uses a mini DHCP server that assigns itself the IP address of 10.0.0.1 and supports 20 clients using addresses of 10.0.0.11 through 10.0.0.30. You see the main login screen. Configure your connecting workstation to automatically obtain an address using DHCP or manually assign it one between 10.0.0.2 and 10.0.0.10.

3. **Enter a password of Cisco using case-sensitive typing.**

   You can tab over the username field. Press Enter. A Summary Status Page appears showing you the current status of the device and a menu.

4. **Select Express Setup.**

   You see a dialog box containing numerous items including SSID, IP addressing, and other items. For initial connections, you can use most of the defaults supplied. However, notice the two fields called Radio0-802.11B and Radio1-802.11A. This device permits connects using either of these protocols, allowing you to use separate connection details for each of them. We provide you details of each setting in Table 5-1. Set your access point according to our suggestions in the table. Click Apply to set them.

Table 5-1	Cisco Aironet 1200 Express Setup
Item	*Description*
System Name	The System name. It appears in the management pages
MAC Address	Shows current MAC address of the device.
Configuration Server Protocol	Indicates whether DHCP or static IP is in use.
IP Address	Negotiated by DHCP or a static address you enter.
IP Subnet Mask	Negotiated by DHCP or a static mask you enter.
Default Gateway	Negotiated by DHCP or a static address you enter.
SNMP Community	The default SNMP community string or password.
SSID	The default is tsunami.
Broadcast SSID in Beacon	Indicates whether the SSID gets broadcast on the network When set to Yes, allows clients to associate without specifying a SSID.

Item	Description
Role in Radio Network	Identifies the role of the device on the wireless network.
Optimize Radio Network For	Used to maximize the data volume or range of the device.
Aironet Extensions	Used when only Aironet devices are on the network.

The last five are duplicated for the 802.11a network. Of the settings described in Table 5-1, determine whether you plan to use a static IP address range and subnet and key those in the applicable fields, setting the Configuration Server Protocol to reflect the change to Static. This Cisco device allows access using Simple Network Management Protocol (SNMP). Make sure that you use a strong private community string here. The community string is comparable to a password, and anyone knowing it can manage the access point — so make sure that it is a strong password.

You can leave most items on their default value, but choose a new SSID in each Radio setting as we mention earlier in the chapter. Note that on the 1200 model, you can set 16 different SSIDs in each Radio, allowing you to set different configurations for each of the 16 SSIDs. This is useful in large networks where you are trying to separate network traffic. There are a couple of new items on this device:

- The **Broadcast SSID in Beacon** option is similar to the SMC option ESSID Broadcast. However, if you add multiple SSID names in each Radio, this option is no longer available.

- The **Role in Radio Network** option allows you to choose between a normal access point and a repeater. The default assumes that it is an access point so leaving it there is usually the best bet in this early stage.

- The **Optimize Radio Network For** field allows you to fine tune the access point's ability to manage traffic. The default handles it automatically. The other options available include

  - *Throughput,* which allows larger data volumes but limits the range

  - *Range,* which produces the opposite by allowing farther range but less data

  - A *Custom setting,* in which you can configure more detailed settings on a special configuration page of the Aironet device

- The **Aironet Extensions** option has a default of Enable. Unless you plan to use only Cisco devices, you should set this to Disable.

Cisco devices include a great deal of additional options because they are intended for Enterprise implementations. Be sure to read your manual before adding one to your network to understand the additional data that we cannot cover here.

Choosing a fail-safe password

Do not use any word in the dictionary. Any dictionary. Such passwords are probably the easiest to attack because all you need is a brute force attacking tool and a dictionary. This book doesn't have sufficient pages to go into the details of performing such attacks, but numerous tools are available because your primary access to the logon process is through a Web page. That is why you don't want to allow remote access — doing so makes it too easy for hackers to attack the device.

So what should you do? Use a combination of alpha (a,b,c) and numeric (1,2,3) characters. If your device permits special characters like @!$, you can combine these also to create an even stronger password. Although you can get carried away figuring out different passwords, remember that you need this password to access the device in order to administer it. The good news is that if you forget the password, all you need to do is physically reset the device, and it reverts to its original settings. Check your owner's manual to learn how this is done.

These devices allow for command line interaction. You do this using a program called Telnet or Secure Shell. We do not recommend using Telnet because the authentication password travels across the network in clear text and is therefore available to anyone with access to the network and a sniffer. Secure Shell is an encrypted method of access. Unfortunately, setup and use of SSH takes more space than we are allowed for this book. You can read all about SSH at www.ssh.com.

Continue using the browser to finish creating a static WEP key. To set up WEP, go to the Security page and follow the prompts. Your access point is now available for use on your network. Connect it as you would the SMC and begin enjoying the wonderful world of wireless.

Testing the Signal

Before you allow your users to connect, you need to make sure that the placement of your wireless networking components remains effective. In order to test the network, you need a wireless laptop configured to the parameters you just enabled. You may want to jump to the next chapter and find out how to set up your machine and then go to Chapter 15 for methods you can use to test the connection. You can perform a rudimentary test using the ping command that we show in Figure 5-3.

Chapter 6

Connecting Your Clients

• •

• •

*Y*ou have an access point installed, but it is of little or no value without clients. You may have Linux or Mac OS clients. Also, you may have Windows 2000 Professional or XP Professional clients. This chapter provides some help on selecting the right hardware and installing the accompanying software. Some of the steps, such as acquiring and installing hardware, are the same regardless of the platform that you choose to use. But when it comes to configuring the software, you need to be aware of some subtleties for each platform.

Generally, connecting your client to a wireless network is a cinch, especially when you are using a wireless-aware operating system like Windows XP, a chipset like Centrino, or a computer equipped with an Intel Pro/Wireless adapter. You will find it marginally more difficult for other platforms; where you have to use the client utility that comes with your wireless adapter.

Of course, you need to read the vendor's documentation for your particular software or hardware, but usually the first step is to install the hardware and to configure your wireless adapter.

Adding Hardware to the PC

The first step when hooking up a client is to install the hardware. But as we spelled out in Chapter 2, you need to plan for the needs of your clients. There are many different forms that the hardware can take, and each one has a niche.

When you complete your site survey, you document the types and locations of all clients. Your clients may include desktops, laptops, tablets, personal digital assistants, and printers. Basically, your client is any device that has an 802.11 interface to the wireless medium. Your Wi-Fi adapter has two major functions:

- ✔ **Microwave software-controlled radio:** This function handles the physical layer microwave reception and transmission, which includes modulation and frequency control. This is the Physical layer.

- ✔ **Media access control and logical link control:** This function enables Ethernet networking over the radio system. It bridges the wired Ethernet network to the wireless 802.11 network. This is the Data Link layer.

Figure 6-1 shows the components in a depiction of OSI layers 1 and 2.

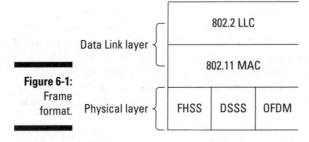

Figure 6-1:
Frame
format.

Look for the Wi-Fi Certified seal on any hardware you purchase. This Good Housekeeping Seal of Approval gives you confidence that your gear will interoperate with any other gear that also has the Wi-Fi certification. You can find a current list of certified gear at www.wi-fi.com/OpenSection/ certified_products.asp?TID=2. Our search on the Wi-Fi Alliance Web site for a CompactFlash 802.11g-certified product found the ARtem CPS-BR-g product. You can look up the equipment you are looking to buy or use this site to narrow down products to evaluate.

Peripheral Component Interconnect (PCI)

Originally, client adapters came in two flavors. You had your PC card (www.pcmcia.org), which some people still refer to as a PCMCIA (Personal Computer Memory Card International Association) card, and your Compact Flash (CF).

In the old days (perhaps reaching back as far as 2002), things were different. When you had a desktop machine, you installed a card in an empty slot and slipped a PC Card into it. Now you can buy a Peripheral Component Interconnect (PCI) adapter with a rubber-ducky antenna. Usually, you can remove the attached antenna and use an external antenna of your choice, such as a semi-directional antenna. All new desktops have at least one PCI slot (usually four or five PCI slots) but rarely (if ever) have a PC card slot.

PC Card adapters use a Type II slot, which is 5.5mm thick. PC Cards have a bulge antenna enclosed in a casing that is thicker than a typical Type II card. The antenna generally blocks the other Type II slot. If you need that other PC Card slot, you may have to go with a USB solution. However, the big problem with PC Cards is the antenna is on a horizontal plane, but most access points have vertical antennae. This causes what is known as *cross polarization,* which reduces the range and the bit rate of the client. Antennae work best when oriented vertically. Really, they work optimally when at a 90-degree angle to the ground. The patch antenna in your PC Card is oriented horizontally unless you turn your laptop on one end — great for reception, but awfully poor for typing.

Some PC Cards are better than others, so make sure that your selection provides the reception your clients want and deserve. Asanté has a card (the FriendlyNet AeroLAN AL1511 PCMCIA adapter) that has a pair of hinged antenna that stows away but unfolds vertically for use. In addition, some vendors such as Alvarion and ORiNOCO have a small plug on the end of the card bulge that you can pop out to attach an external antenna. Last, where reception is a problem, there is another solution. Whereas most client adapters are rated about 15 dBm or 32 milliwatt (mW), you can purchase PC Cards with power as high as 23 dBm or 200 mW to compensate for the bulge antenna. However, you may want to substitute an USB adapter, with its vertical antennae, and move the adapter closer to the access point. Regardless of what Tim the Toolman says, more power is not better. Some low-end microwave ovens use only about 500 mW. Makes you think!

For handhelds in the old days, you needed a CF card. Now not only do you have the CF card, but you also have other flash memory products such as Memory Stick (`www.memorystick.com/en/index.html`) and Secure Digital (`www.sdcard.org`). CF cards, SD, and MS allow you to connect a Pocket PC or PalmOS device to your WLAN. You want versatility; at least one vendor sells an adapter for the CF card to convert it into a PC Card.

Universal Serial Bus (USB)

Not long ago, laptops and desktop PCs did not support Universal Serial Bus nor the USB format. Now you have difficulty finding one that doesn't. First popularized by the Apple iMac, USB format is now ubiquitous. Starting after Windows 95 and NT, Microsoft built in support for USB. It simplifies the connection of one device to another device, and most operating systems now have plug-and-play (PnP) support for USB. So correspondingly, manufacturers started to sell USB models. At first, they placed a CF card in a housing and put a USB interface on it. Figure 6-2 shows a typical USB adapter with a vertical antenna and USB cable. Now they have USB models that are similar in form to a USB flash drive. Figure 6-3 illustrates the USB and other form factors.

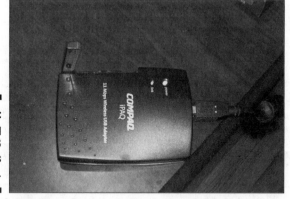

Figure 6-2:
Compaq
USB
wireless
adapter.

Ethernet client adapter

A new entry into the client solution set is the Ethernet client adapter. A good example is the Linksys WET11 Ethernet Client Adapter. These are external devices like the USB gear. They are similar in appearance and footprint to the USB devices. The difference is the device connects to the client using a CAT5 cable and the RJ-45 jack on the computer. You can usually put up to ten feet of cable between the client and Ethernet adapter. So far, most people are using these adapters to wirelessly connect game consoles like PlayStation 2 or Xbox, but you can use it to connect a laptop, desktop, tablet, printer, and any other device that supports Ethernet.

MS SD IO

PC Card

CF card

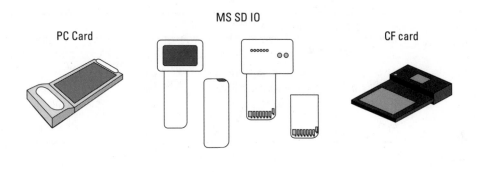

Figure 6-3:
USB
Wireless
and other
form factors.

Mini-PCI   PCI adapter   USB devices

The final decision

Choices, choices, choices. You did do your site survey upfront, did you not?
Table 6-1 lists some of the pros and cons of the various adapters.

Table 6-1	Pros and Cons of Adapters	
Adapter	**Pros**	**Cons**
PC Card	Inside the computer Easy install	Horizontal antenna and poor reception. Bulge antenna blocks other Type II slot. High-power cards eat batteries.
PCI	Inside your computer Flexible, external antenna Can share a PC card with another device	Require you to open computer; not flexible Expensive
USB	Inexpensive Easy install Give flexibility; unplug here, plug in there Power from the USB port Connecting wire allows you to move the device Small footprint	USB 1.1 probably not ready for 802.11a and g. Few have dual or removable antenna. Connecting wire may become a nuisance.

(continued)

Table 6-1 *(continued)*

Adapter	Pros	Cons
Ethernet client	Connect game consoles Act as a wireless bridge Plug-and-play (usually) Connecting wire allows you to move the device	Not common. Expensive. Require external power source.

You may not find the form factor for the standard you intend to support. For instance, you may find it difficult to find USB 2.0 802.11a devices. However, every vendor has different products, different target markets and different strategies, so you really need to talk to your manufacturer or value-added reseller. (Is that an oxymoron?)

Wireless print server

One product also deserving of a look is the wireless print server. The D-Link DP-311P is a good example of a wireless print server with one IEEE 1284 bidirectional parallel port. You can use the print server to share a printer wirelessly. The print server plugs directly into the back of your printer. Figure 6-4 shows an older HP LaserJet 4M with an attached DP-311P. This printer did not support networking and was locally attached using a serial cable. Now the printer is available for sharing through the access point — a great solution for small and home offices.

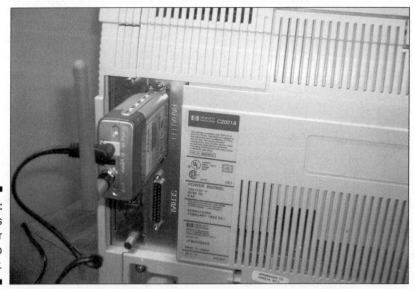

Figure 6-4:
Wireless print server attached to a printer.

Installing the Wireless Hardware

The steps to install your hardware are pretty straightforward. Your vendor may deviate from the following list, however, so you probably should review their documentation after you understand the basic steps.

1. **Insert your CD-ROM.**

   Most vendors want you to install the drivers for your adapter before even inserting or plugging it in. You should find drivers on the CD-ROM that came with your adapter. Usually, you just run the Setup program on the physical media that came with your adapter.

   Should you find you don't have drivers or an install utility, check out the download sections of your vendor's Web site. You have a partial list above. Operating systems such as Windows XP have many of the drivers built in.

2. **Run the setup program and answer the questions with the information you gathered before starting the install.**

3. **Insert, plug in, install, or connect the actual adapter.**

   Your vendor may require you to shut down the device before actually installing the physical hardware. If so, do so and power back up. Often, you will insert the PC Card when instructed to do so by the setup program. Typically, you don't need to power down your system to install a PC Card or USB device, but you will need to shut down when you install a PCI or internal card of any kind. You have the added concern of static electricity for internal cards, but that is not a concern when adding a PC Card or USB device.

4. **If everything went well, your operating system should recognize your hardware.**

   In Windows 2000 or XP, you should see a newly created entry in the Network Connections control panel identifying the wireless adapter. You may have to use the Found New Hardware Wizard to find the drivers for your device.

5. **Upgrade the firmware or software that comes with your hardware.**

Upgrading the firmware or software

Most wireless manufacturers implement features such as security in firmware on the adapter itself. In addition, almost all use flash-able adapters. Periodically, your vendor will post software and firmware updates on their Web site. It is your responsibility to check the site from time to time and download the update. Sometimes, the vendor labels the update as mandatory because it fixes major bugs. Other times, the vendor includes an optional feature that you may or may not use.

We frequently caution people not to install any patch or upgrade without doing some analysis. Just because the vendor thinks the upgrade is mandatory doesn't mean that you must put it in. However, it does mean that you should evaluate the upgrade. You may not use the feature, so why upgrade?

After you download the software upgrade, you just need to unzip and run the setup program. Pretty simple. Be aware that when you upgrade, you lose your connection. You may have to restart your system to reacquire an IP address and to reconnect.

You may find that you are not so lucky. You may have to run a routine that writes new instructions to the chipset. This is known as *flashing the firmware*. Flashing allows you to get new instructions (read as new or enhanced capabilities) on the adapter without having to purchase a new adapter or having to return it to the vendor. Be warned that there is a risk in attempting to flash your firmware. You could lose power to your system and only partially update the firmware, which may also erase the original factory settings — a potentially catastrophic situation. With other vendors, the vendor may allow you to force a reset to the original factory settings — an inconvenience but not a catastrophe.

When flashing your card, we suggest that you use the mains, not battery power. Better yet, plug your computer into a UPS with standby power and plug the UPS into the mains. You don't want to lose power when flashing your adapter!

Some upgrades are significant. When Peter installed his Proxim adapter, his wireless adapter supported no security, pre-shared keys, LEAP (Lightweight Extensible Authentication Protocol), and externally managed 802.1x keys. You can see these features in Figure 6-5.

After downloading the upgrade and installing it, the card has many new features. Figure 6-6 shows the new security tab for this adapter. You can see the client now supports WPA (either shared-key or EAP versions) and more EAP versions. We cover the advantages of these new features in Chapter 11.

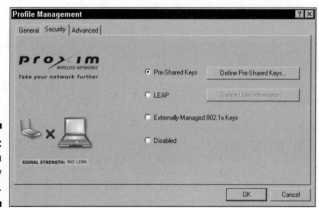

Figure 6-5:
Proxim
Client Utility
Version 2.3.

Figure 6-6:
Proxim
Client Utility
Version 2.4.

Important guidelines for upgrading

Whenever you intend to do an upgrade, follow these simple guidelines:

✔ **Copy your current configuration.** Some vendors allow you to backup the configuration to a file; others do not. If you can make the backup, do so; otherwise, capture all the information from your adapter. You may want to take screenshots. The reason you want to take a backup is that most likely flashing will erase everything you entered and take you back to the factory settings. So you will need to re-enter any pre-shared keys and profiles of sites you visit.

✔ **Read the README file and follow the instructions carefully.**

✔ **Do not turn the power for your client off or unplug anything (especially the adapter) until the upgrade completes.** If you do, pray you followed the first guideline.

Perform the flash over a wired segment or USB connection and not a wireless one. This is more important when doing a flash of an access point because most assuredly, you will lose your connection to the access point when using wireless.

✔ **Plan for and test your upgrade.** Ensure that you can roll back the upgrade should you have the need.

✔ **If you are upgrading more than one client, do one client and make sure that you can still reconnect.** After you determine that the upgrade does not prevent your client from connecting, then by all means upgrade the remaining clients, but not beforehand.

Better safe than sorry. Following the guidelines will take a little extra time upfront, but it may save you hours on the back end.

Configuring the Client's Operating System

After you install and configure your hardware, you need to configure the software to connect to the access point. You need the access point specific information for your organization to setup your clients. At a minimum, you will need the following:

- Choice of ad hoc or infrastructure mode
- SSID (Service Set Identifier) of network where you want the client to connect
- WEP (Wired Equivalent Privacy) encryption keys
- Other keying material for WPA (Wi-Fi Protected Access) or AES (Advanced Encryption Standard)

Configuring Windows XP Professional clients

If you decide not to use the security features, we recommend that you jump ahead to Chapters 9–11 and re-think that decision. But if you don't use the security features, you may find with XP that you are already connected to any open system. Windows XP is the first version of Windows designed to work with wireless networking. Microsoft integrated wireless support into XP for many adapters and provide their associated drivers. Not one to miss a marketing opportunity, Microsoft has coined this *Wireless Zero Configuration*. Zero configuration is probably another example of Microsoft hyperbole, but it is definitely easier (assuming everything works — I am not sure debugging is easier). In fact, you may have a wireless connection and not even know it.

Move your cursor over the Wireless Network Connection icon in the system tray. If the pop-up shows a network name along with the speed and signal strength, you're connected and need do nothing more. Make sure it's your network, or you may have to disconnect!

When another SSID shows up or you're not connected anywhere, you will need to establish a connection with that access point. With Windows XP, you do this with the following simple steps:

1. **Right-click the Wireless Network Connection icon in the System Tray and select View Available Wireless Networks from the contextual menu.**

2. **From the Wireless Network Connection dialog box, select the network where you want to connect.**

3. **Click the Advanced button.**

4. **From the Wireless Network Connection Properties dialog box, click the Add button shown under the Preferred Networks box.**

    You should see the configuration dialog (see Figure 6-7) for a new Wi-Fi network.

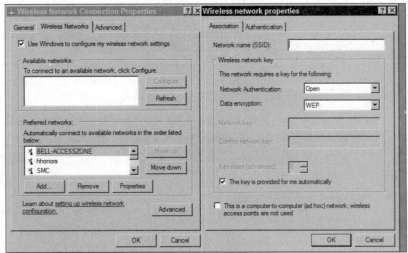

Figure 6-7:
Wireless
network
properties.

5. **Enter the SSID for the network in the Network name text box.**

6. **If you are using WEP, check the Data Encryption box and clear the The Key is Provided for Me Automatically check box. Check the Network Authentication (Shared Mode) box. Enter your first WEP key, and then confirm it a second time.**

    If you are using WPA, choose Shared from the Network Authentication pop-up menu. Check the Data Encryption box and clear The Key is Provided for Me Automatically check box. Check the Network Authentication (Shared Mode) box. Enter your first WEP key, and then confirm it a second time.

    If you are using WPA-PSK, choose Shared from the Network Authentication pop-up menu. Select TKIP or AES from the Data Encryption box. Enter your key in the Network Key field, and then confirm it a second time.

7. **Click the Connect button.**

8. **Repeat steps for every network you want to connect to on a regular basis.**

Configuring Windows 2000 clients

When installing on Windows 2000, you may want to have your OS CD-ROM handy. During the installation process, Windows may need to copy some files from the CD (or from the \Windows\Options\Cabs folder on the hard drive).

Windows 2000 doesn't support Wireless Zero Configuration, but it is pretty straightforward anyway. Generally, you will use the setup program that comes with your adapter. We provide two examples for you in this book: Proxim Client Utility and Boingo.

Configuring the Proxim Client utility

A popular wireless adapter is from Proxim or ORiNOCO. The card comes in various flavors. You will find Silver and Gold models and a/b/g versions as well. To configure the Proxim ORiNOCO a/b/g ComboCard, follow these steps:

1. **Double-click the Proxim Client Utility (PCU) tray icon.**
2. **Click the Profile Management tab, and click the New or Modify button.**
3. **Enter a profile name.**
4. **Enter the SSID of the network in SSID1.**
5. **Click the Security tab.**

   If you are using 802.1x (TLS), you will need a certificate from your server. Follow these steps:

   a. Select 802.1x and EAP-type TLS.

   b. Click Configure and enter the Server/Domain Name and the Login Name.

   c. Click OK.

   If you are using 802.1x (PEAP):

   a. Select 802.1x and EAP-type PEAP.

   b. Click Configure and choose <Any> for the Server Properties.

   c. Enter the User Name and Password.

   d. Click Advanced Configuration. If you did not obtain a certificate from the server, do not mark the Specific Server or Domain check box.

e. Check the box Login Name and enter the Login Name.

f. Click OK twice.

If you are using WPA (TLS), you will need a certificate from your server. Follow these steps:

a. Select WPA and EAP-type TLS.

b. Click Configure and enter the Server/Domain Name and the Login Name.

c. Click OK.

If you are using WPA (PEAP):

a. Select WPA and EAP-type PEAP.

b. Click Configure and choose <Any> for the Server Properties.

c. Enter the User Name and Password.

d. Click Advanced Configuration. If you did not obtain a certificate from the server, do not mark the Specific Server or Domain check box.

e. Check the box Login Name and enter the Login Name.

f. Click OK twice.

If you are using WPA-PSK:

a. Select WPA-PSK.

b. Click Configure and enter the WPA Passphrase.

c. Click OK.

If you are using WEP:

a. Select Pre-Shared Key.

b. Click Configure and enter WEP Keys.

c. Click OK.

If you are using No Encryption, select None.

6. **Click OK.**

7. **Activate your new profile.**

That's it. The Proxim Client utility will work for Proxim adapters on a variety of platforms. For example, you can choose to use the PCU over the Network Connection facility in Windows XP.

Using Boingo to access the network

Boingo is the brainchild of Sky Dayton, EarthLink's founder. Sky aggregated ISPs to create EarthLink. Now, Sky is trying to do the same with WISPs (wireless Internet service providers). Boingo provides a fee-based service and acts as a wireless aggregator, but that's not why it is mentioned here. The brains behind Boingo decided to distribute a free wireless network client that will detect and connect to 802.11 networks. If you are accessing one of Boingo's fee-for-service networks, you don't need to enter any network and security information. The client software manages your user profiles and provides a VPN connection to their data network.

You can use the Boingo client to do network discovery. After you find an available network, you click it and then click the Connect button. Boingo will connect you to any unencrypted access point. In Figure 6-8, you can see a B in the system tray, which indicates that Boingo is installed. You can enter information about the SSID and encryption. To add a profile for your network, just follow these steps:

1. **Click Add to open the Profile Editor dialog, which gives you four tabs.**

2. **In the Network (SSID) field, enter the name of the network or click Browse to see broadcasting networks. If your network is closed, select Does Not Broadcast Its SSID.**

3. **If you use WEP, click the WEP Key tab, check WEP encryption, and select the I provide the WEP Key Data radio button. Enter the key in the Data field. If you want to choose between ASCII and hexadecimal keys, check Advanced.**

4. **Click the Auto Connect tab, and then select how you want the profile activated: as an option (Offer), automatically when you're not connected (Connect), or automatically even when you're connected to another network (Switch).**

    You can also have Boingo run a program, such as a browser, after connecting.

5. **From the IP Settings tab, choose between a static or dynamic IP address.**

6. **Click OK.**

7. **Select a profile and click the up or down Order arrows to choose the order of the auto-connect options.**

Boingo icon

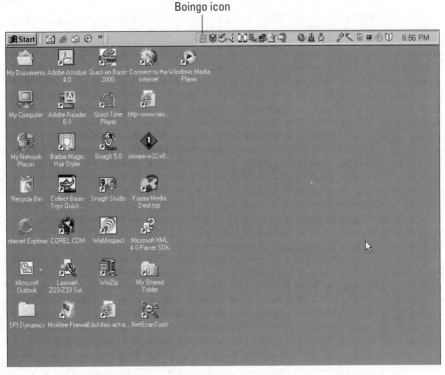

Figure 6-8:
Look for
Boingo in
the system
tray after
installation.

It is a handy little utility to connect to networks, and you can download it from www.boingo.com/download.html.

Configuring Mac OS clients

Mac OS X systems will connect seamlessly and automatically to any access point that uses neither encryption nor Apple's AirPort security. If you find you have access to the network, then you need do nothing more. In the event that you didn't connect or you want to connect to a closed network or a non-Apple access point (that is, not AirPort) with WEP or WPA, you need to use the Intermediate Connection program. For Mac OS X, you follow these simple steps:

1. **From the AirPort menu in the menu bar, choose your network.**

2. **If a password is requested, enter your AirPort password in the Enter Password dialog box, select check Add to Keychain, and then click OK.**

The key to this is in Step 1. It states choose your network from the AirPort menu. This assumes that it is in the menu. If the network that you want to connect to doesn't appear in your control strip or the AirPort menu, you need to add the network. To connect to a closed network, follow these steps:

1. **From the AirPort menu, choose Other to open the Closed Network dialog box.**

2. **Enter the SSID in the Closed Network Name box.**

3. **If you require a pre-shared key or password, choose the key or password type from the Password popup menu and enter the value.**

4. **Click OK to establish the connection.**

Configuring your Centrino systems

You can buy tablet PCs with or without Centrino chips. Centrino chips come in laptops, as well; they are not exclusive to tablets. Intel allows manufacturers to use the Centrino name when the system includes a Pentium M processor, an Intel Pro/Wireless 2100 mini-PCI or similar adapter, and Intel's support chips. If you are interested in Centrino, you can find more information by starting at www.intel.com/products/mobiletechnology.

To configure Centrino, you can use the ConfigFree software, which manages your 802.11, infrared, and wired connections. You can skip ConfigFree altogether should you decide to use Windows XP's network connection facility. Follow these steps to configure your Centrino system using the ConfigFree utility:

1. **Double-click the ConfigFree icon in the system tray or select it from the programs on the Start menu.**

2. **From the hierarchical Profiles menu, choose Open to display the Profile Settings tab.**

3. **Click the Add button to display the Add Profile dialog box.**

4. **Enter a name for the profile and select the appropriate settings (Internet Settings, Devices, and TCP/IP Settings) to include.**

5. **Click OK.**

6. **To use the newly created profile, choose it from the ConfigFree pop-up menu or select it from the Profile Settings tab and then click the Switch button.**

Centrino is very wireless-aware, so it is easy to create new profiles and to switch back and forth between profiles. Centrino-based systems, like Windows XP-based systems, will try to connect to access points automatically. This may cause you some problems when the site is not yours, and they don't require any security.

Configuring Linux and FreeBSD Clients

If you are reading this section, you probably realize that Linux and FreeBSD are arcane. Many great similarities exist between Linux and FreeBSD versions, but there also are many differences and subtleties among them. Fortunately, the Linux and FreeBSD communities have rushed in to support Wi-Fi. To skip Linux and FreeBSD from our discussion is not right because these platforms are making inroads in some organizations. Frustrated with Wintel, these organizations are starting to implement Linux workstations. To attempt to show steps for all Linux (and for that matter FreeBSD) variants is a daunting and most likely fruitless task. There is help, nevertheless.

First, vendors such as Proxim include an archive file with driver source files and other related files on the installation CD-ROM. If you can't find that on your CD, don't fret. You can get general Wi-Fi support for many Linux and FreeBSD variants. For example, check out `www.hpl.hp.com/personal/ Jean_Tourrilhes/Linux/Wireless.html`. You will find links for drivers and tools on this Web site.

If you have an Intersil card based on the Prism chipset, you should check out `www.linux-wlan.com/linux-wlan`. You will find drivers for many Linux distributions. If you have a Prism chipset and FreeBSD, check out `www.freebsd.org/cgi/man.cgi/?query=wi`.

Don't know whether you have a Prism chipset or Lucent chipset? The following manufacturers use the Prism (Intersil) chipset:

3Com	Addtron	AiroNet
Bromax	Compaq WL100	D-Link
Farallon	GemTek	Intel
LeArtery Solutions	Linksys	Netgear
Nokia	Nortel	Samsung
Senao	Siemens	SMC
Symbol	Z-Com	Zoom Technologies

If you have a Prism (Intersil) chipset you will see the computer with antenna icon in the System Tray. Figure 6-9 shows the computer with antenna icon.

Prism icon

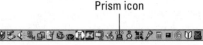

Figure 6-9:
Prism icon.

That's quite an exhaustive list. But quite a few use the Hermes (Lucent) chipset as this list shows:

1stWave	Agere/ORiNOCO/Proxim
Alvarion	Apple
ARtem	Avaya
Buffalo	Cabletron
Compaq WL110	Dell
ELSA	Enterasys
HP	IBM
SONY	Toshiba

If you have a Hermes (Lucent) chipset you will see the step icon in the System Tray. Figure 6-10 shows the step icon.

Hermes icon

Figure 6-10:
Hermes
icon.

To find information for your Hermes chipset, revisit www.hpl.hp.com/ personal/Jean_Tourrilhes/Linux/Wireless.html and search on Orinoco.

Making Sure the Connection Works

Now all you have to do are some tests to determine whether your wireless adapter works. First, choose Start⇨Run and enter **command** in the Open box. A window opens and put you at the command prompt (or because old habits die hard, the DOS prompt). At the command prompt, enter **ipconfig /all**. You should see your wireless adapter as one of the interfaces presented. One of the entries associated with the device is the IP address. Note this address. From the command prompt, ping the address you just noted. If you don't get a successful reply, something is wrong with your adapter. If this works, try pinging another IP address for another system. If this works, then away you go to check out www.pdaconsulting.com or www.cerberus-isc.com.

Chapter 7

Building the Multi-Zone Network

· ·

· ·

Are you in the right zone? If you are a runner or other serious athlete, you know all about zones. Getting into the right zone increases your chances for success at whatever you do. In wireless networks, multiple zones might allude to the number of locations where you provide wireless access or commercial hot spots that advertise themselves as *Wireless Zones*. This chapter shows how connecting to multiple locations seamlessly is the goal, whether it is using a Wireless Metropolitan Area Network (WMAN) like that in Fredericton, New Brunswick, or multiple access points in your own network.

Roaming Around with a Wireless Machine

Ah, the joys of wireless access anywhere in your business. Imagine just wandering around with your tablet PC or PDA and connecting to the Internet or receiving e-mail as you walk the corridors of power. All well and good, but if you are not properly prepared, that dream can turn into a nightmare. Remaining connected as you travel across multiple locations requires careful consideration, planning, and skill.

It is easy for the user to use wireless services and complain about glitches, but what if you want to implement such roaming in your company? You need to know what steps to take, and we show you in the next few pages.

Get in the zone!

With your PDA or other accessory, you can find the closest Wi-Fi Zone by adding the URL `wap.wifizone.org` to your cell phone browser and going there when you need to locate wireless access points. Accessing this site downloads a small WAP file to the phone that sets up your connection. After selecting the site, you can scroll through country, state, and then local access points in each city. Staying connected while roaming has never been easier. This particular service is provided by the Wi-Fi Alliance. Although users are not charged a fee to locate places they can connect, actual connection fees vary depending on the provider. Barry used the service to locate wireless access points near him and found a number around Toronto. He will check it when he gets to Seattle to find access points there so he can possibly connect for free and get his e-mail.

Wireless roaming standards

When you implement a wireless network, you must consider national roaming. We provide some guidance on where current efforts appear to be headed and then focus on what you can do now to achieve roaming for your business users.

Currently, it is difficult and expensive to connect between vendors' wireless networks. This is less because of technical reasons than pure marketing and cost-sharing issues. Vendors such as T-Mobile and Boingo are starting to offer reciprocity agreements, but there is a long way to go for a truly transparent connection like what you typically enjoy with your cell phone.

Asia and Europe are a bit farther ahead than North America with companies like Monzune (`www.monzune.net`) and NTT DoCoMo (`www.nttdocomo.com/`) pushing for a standard for all to follow. It isn't truly necessary to have one standard for all, but doing so will certainly make the transition to wireless easier. In Japan, the Multimedia Mobile Access Communication Systems group is creating a standard for carriers to follow. Its intent is to enable transmission of ultra high-speed, high-quality multimedia information "anytime and anywhere," including providing seamless connections to optical fiber networks. This is a true wireless access vision!

Some of the standards needed worldwide include seamless connectivity, robust single-point payment schemes, automatic network prioritization, and support for push applications — meaning static IP addresses. Current standards for roaming are not fast enough to support applications like voice or multimedia.

IEEE forms a standard

The Institute of Electrical and Electronics Engineers (IEEE), a worldwide standards group, has produced a formal task group (P802 Handoff Study Group) to create a specification for handing off signals as users move from one access point to another. The companies Airespace (www.airespace.com), Cisco (www.cisco.com), SpectraLink (www.spectralink.com/), and Symbol (www.symbol.com) are apparently signed on to help the effort.

You can locate the IEEE 802.11 P802 Handoff Study Group at http://grouper. ieee.org/groups/802/11/index.html. This location even has pictures and e-mail address information for all the members. Wow. Talk about getting junk mail.

In general, there are three phases in the communication process between the access point and a station. Your laptop is able to use these features to move between access points.

- **Network discovery and selection:** Network discovery locates wireless networks for you, allowing you to choose the one you want to use.

- **Authentication/association:** Associations are dynamic in nature because stations move and are turned on and off. A station can be associated with only one access point. This ensures that the Distribution System always knows where the station is.

- **Disassociation/reassociation:** Association supports no-transition mobility, but not enough to support BSS-transition. So this is where reassociation comes into play. This service allows the station to switch its association from one access point to another. Both association and reassociation are initiated by the station and not by the access point.

   Disassociation is when the association between the station and the access point is terminated. This can be initiated by either of the devices.

You can find out all about the IEEE wireless effort at http://standards. ieee.org/wireless/overview.html, where they promulgate all their efforts. Because this in the early stages, no obvious solution is being touted from the IEEE at this point. A workgroup is set up to handle it, however, and define a standard so that interoperability will increase.

Canadians form a standard

A number of Canadian wireless carriers have signed an intercarrier Wi-Fi roaming agreement. This agreement establishes a common standard for roaming and interoperability between public Wi-Fi hot spots. To facilitate marketing efforts, the carriers created a common brand identifier for Wi-Fi hot spots across Canada.

Bell Mobility, Microcell Solutions (Fido), Rogers AT&T Wireless, and TELUS Mobility will use consistent branding with a common hot spot identifier. Plans exist for other Canadian operators or hot spot owners to join the roaming alliance providing they meet the standards of the group. The group plans to build over 500 hot spots within 2004. Using a common access method from Tatara Systems, Inc. (`www.tatarasystems.com/contentmgr/showdetails.php/id/216`), users will see a similar sign-on screen no matter whose network they connect with and use. The company's product is the Tatara Wi-Fi Service Delivery Platform — quite a mouthful, don't you think? Currently customers must use a credit card and pay in advance for this Wi-Fi access. The service expects to eventually allow charges for Wi-Fi usage to go to the users' respective wireless providers. One bill, no matter where you roam. Now, if that can be extended to include the rest of the world, we'll really be impressed!

Payment options

As you roam around the world using wireless hot spots, you will want an effective and simple payment process. In addition to potential payment by wireless provider, a formal standard for sharing Wi-Fi roaming revenue is expected to be available soon. The Internet Protocol Detail Record Organization (`www.IPDR.org`), an open consortium for interoperable service measurement and exchange, has approved the WLAN Accounting and Settlement (WLANAS) service specification. This is designed to standardize settlement costs among hot spot operators and home service providers. The standard supports native environments for various Wi-Fi providers, including GSM, CDMA, WISP, ISP, and wireline.

According to its Web site, IPDR.org is an open consortium of service providers, software and equipment vendors, system integrators, and billing and mediation vendors — all of whom are collaborating to facilitate service measurement and exchange for next-generation services by implementing *de facto* standards for everyone to follow.

You can get a copy of the WLANAS and other specifications that the IPDR group produces at `www.ipdr.org/working-groups/wlanas.html`, where you can also find a Microsoft PowerPoint presentation to download that describes these efforts. The standard should be undergoing field testing by the time you read this book.

WiMAX standard

The WiMAX standard consists of "the last mile," as it is sometimes referred to in industry. *The last mile* is that gap where a wired network ends and a business or home that wants access to the network resides. This gap is often too costly to fill or the terrain is too difficult to wire. Bridging it with a strong wireless connection allows for remote locations to achieve connectivity. This is the 802.16x wireless metropolitan area network (WMAN) specification, currently being developed and promoted by the Worldwide Interoperability for Microwave Access (WiMAX) industry group, two of whose members are Intel and Nokia. As with Wi-Fi, the term WiMAX is becoming widely accepted as the name for the 802.16x standard.

The WiMAX group takes on a role similar to that of the Wi-Fi Alliance, promoting standards certification and interoperability. The initial standard operates in the 10–66 GHz frequency band and requires line-of-sight towers, but a new extension 802.16a, ratified in March 2003, provides for a lower frequency consisting of 2–11 GHz. This helps ease regulatory issues, and perhaps more importantly does not require line of sight. It boasts a 31-mile range, compared with Wi-Fi's 200–300 yards, and 70 Mbps data transfer rates, making it an important adjunct to that last mile. That isn't the only area for growth of this standard, however. Consider the following:

- Enterprises can use WiMAX instead of T1 for far less cost, and SMEs/SMBs/SOHOs can be offered fractional T1 services.

- Wi-Fi hot spot operators may be able to build a spot cheaply, but then they need to anchor it to the public network, and this is usually accomplished by using T1 or DSL. They might use WiMAX instead.

- Rural areas where there is no wired or cellular infrastructure, nor the incentive to spend on delivery of such service, can now be offered wireless access.

The new standard uses point-to-multipoint, and, as mentioned, does not require line of sight. It uses licensed bands at 3.5 GHz and 10.5 GHz internationally and 2.5–2.7 GHz in the United States, along with unlicensed 2.4 GHz and 5.725–5.825 GHz bands. The IEEE is currently reviewing 802.16d, which is a combination of both the 802.16 and 802.16a standards; following that, the IEEE will probably then approve standard 802.16e, which is more of a roaming-friendly version of 16d.

Expect to see more license-exempt wireless ISPs offering WiMAX service in the near future. Apparently, there are already about 1,800 such wireless Internet service providers (WISPs) in the United States, and that number is growing. Also, you can expect to see vendors begin to offer wireless access cards supporting both Wi-Fi and WiMAX, further easing the connectivity burden for users. For readers interested in starting their own WISP, this is a standard you will want to learn more about. You can do that at www.wimaxforum.org/home, where you can read about the current issues and implementations.

Remember, WiMAX is primarily more of a broadband solution than a Wi-Fi access point type solution, so it is not something that a typical SME/SMB/SOHO will use except as a service from some vendor for connectivity to a larger network. Expect to see a WiMAX Forum Certified™ certification of products sometime late in 2004.

Connectivity issues as you move around

So what happens when you try to roam around and move from one access point to another? When an access point is broadcasting, it sends out a beacon signal to all the nodes. Each node can select the appropriate beacon (and hence, access point) usually based on the strongest signal.

Wi-Fi supports a range of only around 300 feet and is typically less than that with interference factors, so as you move around, the most common problem is falling out of range of one access point before reaching the range of another. In your business, you avoid this scenario by performing a site survey and then implementing according to those findings, as described in Chapter 2.

Sometimes that beacon just doesn't reach you, and your connection drops. Without a signal, the network just isn't going to work. But what if there was a signal available and you just didn't know it? This is where roaming with multiple wireless providers comes into play. Perhaps you really need that connection and are willing to pay for it. You could use a GPRS service when 802.11 is unavailable. To do this, though, you need both a Wi-Fi and a cellular wireless card. After you have these, you can log in as needed.

This may not be ideal, however, because it still means losing the signal and reacquiring all the information you were using. The PCTEL company (www. pctel.com) may have the solution you need, offering a software product called Segue Roaming Client, which manages wireless connections on Windows laptops and Pocket PCs, including Wi-Fi, GPRS, CDMA, Bluetooth, and infrared connections technologies. Now we are talking connectivity. You can see how the types of connections use Segue in Figure 7-1.

How does Segue work?

Segue acts as your interface to the wireless world, informing you of all potential connections and allowing you to select priorities among those found. One great use for this is for a service provider to offer the client software to their users and perform remote configuration and updates through a central configuration server.

In addition to support for a number of wireless networks, the software also provides support for VPN tunnels, WPA, and 802.1x security, as well as a customizable interface.

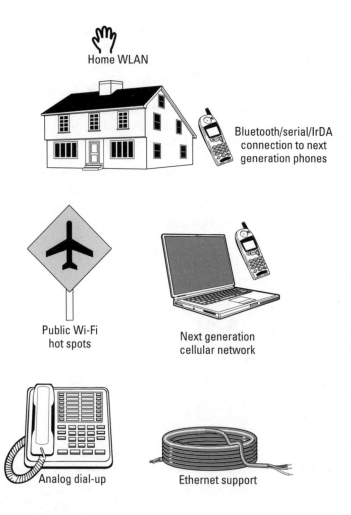

Home WLAN

Bluetooth/serial/IrDA
connection to next
generation phones

Public Wi-Fi
hot spots

Next generation
cellular network

Figure 7-1:
Examples of
wireless
connections
using Segue
Roaming
Client.

Analog dial-up

Ethernet support

Installing Segue

Installing the client is simple. Here's how to do it in a Windows XP system:

1. **Download the software from** `www.pctel.com/prodSegRCdnld.html` **to your hard drive.**

2. **Double-click the downloaded software to start the install. After the installation wizard appears, click Next.**

3. **Select the check box to agree to the license terms and then click Next.**

4. **Select Express and then click Next. Then click Next again to start the install.**

5. **Click Finish to end the installation and open the client software.**

After the client software is loaded and running, it automatically checks your wireless network card and uses it to discover available networks. If you install both a GPRS network card along with your 802.11 card, you will see both network types listed if they are available.

Configuring and managing your connections

The software enables you to configure and manage your connections. Clicking the menu button provides a list of things you can do with the software.

To choose from the available networks, follow these steps:

1. **Select Network List from the menu.**

   You see a picture showing available networks, their names, BSSIDs, Channels, Encryption, and signal strength. Of course, you could use this to roam around looking for wireless networks. You see this clearly in Figure 7-2.

Figure 7-2:
Available
networks
using the
Segue
Roaming
Client.

2. **Click the Connect button for the network you want to use.**

   Note that the picture shows only one viable network; the strength of the other network appears to be minimal. Yours may show different results.

3. **When you connect for the first time, or if you decide to connect to a network that uses encryption, you get a message indicating that it needs a security key, such as WEP or 802.1x. Input the correct credentials so that you can connect.**

   After you enter the correct credentials, you are connected to the network.

4. **When prompted by the software, indicate whether you want to be automatically logged on to the connection in the future or be prompted.**

   After you respond, the software connects you to the wireless network you chose.

The tool offers a number of other facilities and is a useful adjunct to your laptop if you travel and need to stay connected. At a client price of around $20, it's quite a bargain. Of course, Segue is not the only solution available. To some extent, Boingo, Winc (www.cirond.com), and others offer similar services. NetMotion Wireless' NetMotion Mobility (www.nmwco.com/default_swf.asp) is another vendor product that provides seamless roaming between wireless networks for Pocket PC-based devices and Windows machines.

Reassociation — Getting back together as you move from AP to AP

Setting up multiple access points allows you to form a larger wireless network, which should be fairly obvious to you by now. Using your site survey data, locate your access points so that they slightly overlap. Of course, you can also use repeaters and bridges to enlarge your network in addition to merely placing multiple access points. As a user moves around with her wireless device, she will encounter each access point as its signal grows stronger than the one she's currently assigned to and automatically switch to the better signal. This is called *disassociation* and *reassociation*.

Using client software such as the Segue Roaming Client does this association for you automatically even across different wireless networks, using the criteria you supply. Of course, you need to be using the correct SSID and security parameters for each network.

Change the channel

To set up your network using multiple access points, you need to set up a unique channel for each access point, within the range of 1–11 for North America. This allows for up to 11 access points to be set up on one LAN; however, it's recommended that you use increments of three. So with three access points, you might use 1, 6, and 11 as channels. Next, you ensure that the service set identifier (SSID) that identifies your Independent Basic Service Set (IBSS) is the same for each access point, and if WEP or other security is enabled, that they all use the same key. When two or more infrastructure Basic Service Sets (BSSs) use the same SSID, a client is able to roam between those BSS's primarily because the SSID is used to designate an Extended Service Set (ESS), which is a collection of BSS's linked by a distribution system. Now, as a laptop or PDA roams from one part of the network to another, the wireless network card will automatically select the strongest signal, changing as one nears and another falls away.

The wireless city

If you travel a lot and like to remain connected, visit Fredericton, New Brunswick, Canada. The city has installed one of the first WMAN networks, with free (yes, we wrote *free*) access to the Internet from almost anywhere in the city. This is really staying connected while you work, eat, and play. This new implementation allows anyone with wireless connectivity to connect while at the local parks, in restaurants, hotels, or walking down a street. It even has a cute name, Fred e-Zone. They use more than 100 Cisco 802.11g access points across the city, in buildings, on rooftops, and even on traffic lights. With future plans to increase wireless access to police, firefighters, and meter readers, Fredericton is well on its way to becoming one of the first cities to be truly wireless.

RoamAD (www.roamad.com) is a company that supplies such services. RoamAD supplies systems and services for the design and operation of metropolitan Wi-Fi networks. Their metropolitan Wi-Fi network technology allows operators to deliver telco-grade mobile broadband services to metropolitan areas. One of its implementations is Reach Wireless, New Zealand. This firm just implemented a metropolitan Wi-Fi network in Auckland. The Auckland central city hot zone is apparently the largest of its kind. Other such network firms include MeshNetworks (www.meshnetworks.com) and LocustWorld (www.locustworld.com).

Roaming mechanisms use either an active or passive type of scanning to find each access point. With *active scanning,* this typically involves the client device sending probe requests on each channel and waiting for probe responses from an access point. *Passive scanning* listens for beacon frames on each channel and doesn't send out probes. Most access points use an active scanning technique. To find out how your roaming mechanism performs, search the vendor sites or your documentation.

Most of this information assumes you are roaming across the same subnet and not separate ones, which is more difficult to set up and requires more than this introductory book can give you. There are indications that you could use MobilIP. This is a system by which an IP network node may roam throughout various IP network structures and send and receive traffic normally. You can find out more at www.personaltelco.net/index.cgi/MobileIp. There is no access point to our knowledge that allows roaming across networks separated by a router. Basically, your users will need to reconfigure their IP settings whenever they cross subnet boundaries, which is fairly unpalatable. A better solution that is more expensive involves setting up a VLAN for your wireless network and placing all the access points on their own subnet, effectively creating a specific network just for your wireless users.

Use an antenna and amplifier

Of course, if your needs are small enough, then by using the right antenna and amplifiers, you should be able to easily accommodate the lunch room, conference room, and a few other locations where you might need wireless access. Then add an authentication process for the wireless machines before allowing them to access the rest of your network using 802.1x protocols. This way you are in effect setting up a Wireless DMZ.

Vendor concerns

You may find that some vendors increase security within a single access point, but then lose the ability to roam. Your solution needs to provide for roaming. One method might be to use a VPN connection or encryption protocols like IPSec, where the security is in effect from the workstation to the WDMZ's network gateway. This way, the access point being used is transparent, and your roaming is seamless.

When thinking of roaming, consider buying all your equipment from the same vendor to help ensure compatibility.

Load Balancing — Are All Zones Used Equally?

So how can you ensure that one access point isn't overloaded while others are virtually unused? For the really technical among you, visit www.ieee-infocom.org/2001/paper/172.pdf and download a white paper outlining the key aspects. Much of the basic load balancing is accomplished by your device. For instance, D-Link Air Premier devices offer automatic load balancing so busy access points can offload to those with the least traffic or least number of users. Clients in a load-balancing network typically set their network card channel option to automatic so that the access points can flip the user around in order to balance the overall network load.

Continuing with the D-Link example, you might implement three access points as overlapping, each connected to your Ethernet LAN. They need this so they can pass the Inter-Access Point Protocol (IAPP), which is used by the load balancing function. Using the setup we mentioned earlier and this protocol, the access points will automatically provide for load balancing. For typical small wireless networks, this is more than adequate.

Of course, if the number of users or the network load in an area is too much for one access point, install several overlapping each other. Using the same configuration details as we mention earlier (different channels, same SSID), you can increase effective bandwidth in an area using multiple access points near each other.

Load balancing is performed using each vendor's specifications, aside from the simple act of assigning different channels. Whether one is better or more useful to you than another is always a point for discussion. We urge that you review each vendor's materials to make an informed decision.

With no real solutions for load balancing and no specific standards, an Internet Engineering Task Force (IETF) working group has been proposed in the Operations and Management Area to will look at, among other things, load balancing. The working group hopes to be implemented and advise on solutions within the year and is tentatively called the *Control and Provisioning of Wireless Access Points (capwap) working group.*

Chapter 8

Using Wireless on the Road to Connect to the Office

On the road again is a familiar refrain for many of us in the business world as we travel across the country or across the world in search of business and pleasure. Using wireless connectivity helps ease the burden of that travel as we connect to our businesses or e-mail or homes. Barry even sometimes uses network connectivity to connect to home with a camera so he can see his wife and chat with her while whipping around the world. This chapter outlines some of the methods you use to connect to the office while doing all that business travel. Let's face it: Keeping everyone working even while they are away is *de rigueur* these days.

Spontaneous Communities: Ad Hoc Networks

The advent of mobile ad hoc networks (MANET) offers opportunities for spontaneous communication among groups of people who are near each other but perhaps not previously connected in any way. These networks are limited by dynamic changes in the topology and the availability of resources at any given time. Typically, such networks consist of short-range devices using Bluetooth, for example. You can form one merely by being part of a large group of people who have a small-range network component like Bluetooth on some device or other.

As can be expected, such connections are built and torn down dynamically as people enter and leave the area. They might also be called *peer networks*. Expect to see a surge of such networks in the years to come as they catch on, especially with the younger crowd who like to socialize and also are attuned to technology. If such infrastructures are coupled with mobile ad hoc applications, then the possibilities become endless. Imagine using such a network to connect with others of like interest. For instance, imagine attending a technical conference using an application (into which you enter your specific needs or key questions you need answered) that inquires whether like-minded individuals are nearby, thus allowing you to connect to them while you wander through the conference grounds. This could allow much closer interaction with like-minded individuals, allowing for a sharing of ideas and fostering an improved ability to bring back critical information to your firm. Responding, of course, is entirely up to the individual, but it would certainly help you get answers to questions when you are unsure who to ask, wouldn't it?

Likewise, another possibility might entail setting up a number of interests and seeing who shares those interests as you join a local support group. You could then connect wirelessly, share some data on what you look like, and then get together if interested. The possibilities are endless.

Finally, what about setting up a user community for specific technical backgrounds and having a host server with data on it available for access. When everyone gets together, you can share thoughts using wireless and share data by adding or taking it from the server effortlessly.

Wi-Fi Warriors on the Road

Business travel is increasing significantly it seems as more and more companies begin to compete on a global scale. Although few travelers enter the rarified scale of Platinum or Super Elite, with over 100,000 miles a year of travel, more and more mid-range travel is expected to occur. Despite all the supposed benefits, that is a lot of miles and a lot of time away from home.

With all that travel comes the need to remain connected in order to continue your business and connect with family and friends. This is becoming easier to do with the advent of wireless networking. No longer do you need to hunt down a telephone so you can dial in; you can look for a wireless connection in the hotel, restaurant, or airport, and connect from there.

In fact, as you see in Chapter 19, you can even connect with another vehicle using your wireless network card and your friend driving close behind or beside you.

Wireless at the airport

Airports are the traveler's nightmare. Some are huge and sprawling like Chicago or London's Heathrow, and others are small and easy to navigate like Dubai International. One of my favorites is Amsterdam's Schiphol with its simple rail system and close proximity to downtown Amsterdam. In fact, on some of those interminable layovers, you can either visit the small museum and view some of the world's masterpieces at no charge or take a tour of downtown with a travel service that guarantees your return in time for the next segment of your flight. As every traveler knows, airports can be great, or they can be miserable, depending on your point of view.

Getting connected with your laptop can be challenging unless you are privileged enough to travel in Business Class or First Class where lounges are more likely to offer such services. For most travelers, wireless access is only just becoming available. It is worth the cost, though, if you need to be available to provide support or handle other issues as they arise, without needing to wait until you can reach your hotel.

To discover which airports offer wireless, you can visit each one's Web site (an onerous task if you travel a lot), or you can go to several Web sites that provide up-to-date lists, like www.ezgoal.com/hotspots, which also provides hotel and metropolitan hot spot listings. This site provides many pages of listings although most of the airports listed are located in the United States. Some of the airports currently offering wireless access include

- Buffalo Niagara International Airport, USA
- Chicago's O'Hare International, USA
- Domodedovo Airport, Russia
- Minneapolis-St. Paul International Airport, USA
- New York's LaGuardia Airport, USA
- Narita Airport, Japan
- Ottawa International Airport, Canada
- Oslo Airport, Norway
- Pittsburgh International Airport, USA
- San Francisco International Airport, USA
- Schiphol Airport, The Netherlands

One warning about some of these providers though: The daily charge is only good until midnight, so if you arrive late in the day and want to connect, it may end up costing you two days' worth if you remain connected past midnight. Not a very considerate model for international travelers who often arrive late and leave early.

Wireless in hotels

Traveling and hotels are synonymous. A large part of a hotel's revenue comes from the business traveler, who usually needs to conduct business while at the hotel. That's where wireless is becoming useful. The many hotels I have stayed at and worked in, often late into the night meeting deadlines or preparing for the next day's presentation, provided little in the way of network access. Most of the better hotels offer a data port — and if you are lucky, a small business table to work at with easy to reach power connectors. However, in just as many hotels, you fight for a power connection often needing to unplug the lamp in order to power your laptop.

Make sure that you bring a power bar, international power connectors, and a small toolkit when you travel. You'll need them all eventually.

Luckily, the trend is reversing with wireless access leading the way. Choice Hotels International, Inc., which owns 370 Comfort Suites and 140 Clarion properties, plans to roll out free wireless Internet access in public areas and their guest rooms this year. Marriott International has wireless high-speed Internet access at 400 hotels in the United States, the United Kingdom, and Germany. In addition, all Fairmont hotels around the world use wireless access points. The Sheraton at Amsterdam Schiphol airport offers wireless as well as its cousin in Toronto. And so it goes, with more and more hotels joining the bandwagon.

It is estimated that the average cost of a single hotel hotspot deployment is around $2,000, which is far less expensive than hard wiring all the rooms for those hotels with only dial capability.

To locate access points in hotels or coffee shops, Boingo (a wireless vendor that has outlets around the world in hotels and cafes) and a few U.S. airports that use its Web site (www.boingo.com/search.html) allows you to input a country and locate all the hot spots by category. Although this is a tad expensive unless you use hot spots a lot, Boingo does seem to have a good selection of locations around the world

In the Table 8-1, we offer some of the current hotels that provide wireless connectivity and an idea of where connectivity can be found within the hotel. This might help in your travel decisions if you need to remain connected to better serve your organization. Note that not all locations offer connectivity everywhere because it is early in the adoption cycle for some; thus, only a small percentages of their rooms may be connected.

Table 8-1	Hotels with Wireless Connectivity
Hotel	*Locations*
Best Western	Public areas and guest rooms
Fairmont Hotels and Resorts	Public areas, guest rooms, poolside
Hilton	Public areas and some guest rooms
Hyatt	Public areas, some guest rooms
Marriott	Public areas, meeting rooms, some guest rooms
Microtel Inns & Suites	Public areas and all rooms
Starwood Hotels	Public areas, some guest rooms

This is certainly not an exhaustive list and will change quickly as hotels add locations and increase room penetration, but it gives you an idea where they are headed. Call ahead and ask when you begin your travels.

With wireless access in hotels expected to jump to over 38,000 by 2008, it certainly looks like wireless is here to stay.

E-Mailing Wirelessly with Microsoft Exchange

Obtaining your corporate e-mail while traveling is always a good thing, right? Okay, it's a good thing for your parent company because it at least can keep you working. So how can you accomplish that while you zip around the world, with all that free time you have sitting in an airport waiting for a cancelled flight?

We mention in Chapter 12 about using a Virtual Private Network (VPN) to connect to the office, and that is always the best method. It helps ensure that you have full access to all your business applications as well as letting you get mail as if you were in the office. On top of all that, mail is encrypted while it crosses the Internet into your laptop, and that is definitely a good thing.

However, there are other alternatives, such as connecting to a POP account or directly connecting to your corporate e-mail server. In using a POP account, you set up a special account just for travel and have the office forward only important e-mails to you. It's a little sloppier but gets the job done without the need for a VPN or exposing your e-mail server to the outside world. Many organizations have POP connections to their internal e-mail or allow Microsoft Exchange access though so those can be used instead.

Setting up POP access

Setting up a POP account can be easy. You can even use your PDA to do it so that you don't need to lug that laptop around if that is all you carry it for when you travel. Barry recently set up his Treo 600 to access a special POP account so he can obtain e-mails from those persons he provides the account address to allowing instant access to him as long as his phone is on. Using his GSM-based phone, he used the steps that follow. Similar steps can be used on many properly equipped PDAs. To start, you need to ensure that your service provider offers such a service, that you purchase the necessary data services, and they are activated and ready to use. Of course, you need a POP-based e-mail provider, such as AT&T WorldNet or EarthLink. Check with your provider.

1. **Ensure that access is available from your service provider and that you have activated it.**

2. **Choose the Mail icon. (Read your vendor documentation for the precise icon to use.)**

3. **Look for the name of your service provider in the list provided and select it.**

4. **Fill in the dialog boxes as requested with your name, e-mail address, and password.**

5. **Click Finish.**

Access such as this, though, is not for the faint of wallet. Most providers charge by the amount of data in addition to the dollars you are already paying for the telephone service. However, getting your e-mail on an instantly available basis might save you that big contract or get you the big order.

You may find you need to manually enter your service provider information because they are not listed on the PDA. To do this, you need to obtain IP address information from the vendor and enter that manually. Table 8-2 illustrates a setup for AT&T that Barry uses on his Treo 600. Similar settings apply to most vendors with POP e-mail service.

Table 8-2	TREO 600 POP Mail Settings
Field	*Setting*
Account Name	Enter your name here.
E-mail Address	The address you use for this account.
User Name	The username given to you by your service provider.
Password	Select a good password and enter it here.

Field	Setting
Incoming (POP3) Server	Enter the name for your provider, such as `pop.attglobal.net`.
Outgoing (SMTP) Server	Enter the name of the server to send your reply e-mails to such as `smtp.attglobal.net`.
Leave Mail on Server	Select this field so you can also download the mail to your laptop if you choose.
SSL Required for POP	If your service provider offers SSL protection, check this box.
SSL Required for SMTP	If your service provider offers SSL protection, check this box.
Outgoing Mail Server Username	Enter your user name from your service provider.
Password	Enter the password here. These may be different from your POP account. If not, then you can use the same account and password as entered previously.

SSL is the acronym for Secure Sockets Layer and is the technical term for the encryption used to protect your e-mail or your Web pages. You probably see this all the time as you browse the Internet if you notice the little lock icon that is shown whenever a Web page is encrypted in Microsoft Explorer. It is highly advisable to use this for POP mail because those user accounts and passwords are sent over the air in *clear text* — meaning they are susceptible to interception by anyone wanting to take the time and energy to steal them from you.

Of course, many other PDA utilities offer e-mail service. These include SnapperMail (`www.snappermail.com`) or Visto's Goodlink (`www.visto.com`) and the Seven product line (`www.seven.com`), which offer more advanced e-mail options and even the ability to accept and read attachments.

SnapperMail is a tool aimed more at the home user or SMB who needs a solid e-mail system but doesn't necessarily want the cost and management of the more full-fledged Goodlink system. A user license for Snapper is only around $30 depending on the options one purchases. (Prices are in US dollars.) SnapperMail extends your basic PDA e-mail system to new heights with attachment support, readers for JPEG and Zip files, and the ability to store attachments on your memory card. It indicates that it is Exchange- and Lotus Notes-compatible, but it must be pointed out that this access is really only POP support and doesn't allow for anything other than access to your Inbox. IMAP lets you have access to all of your Exchange folders. SnapperMail does indicate it plans to provide IMAP4 support in the near future however, so this may still be a viable tool for those who can wait.

The Goodlink product costs about $330 annually, so it's only for the corporate person who really requires solid, online access his their e-mail, contacts, and other services. To link with Microsoft Exchange, you need to also purchase the Server version, which adds significantly to the bill but provides for a full range of corporate level services, allowing your entire team full access as though they were at the office.

Seven offers similar service to Goodlink, providing several versions of its product starting with a Personal Edition and topping out with its Enterprise Edition. Seven products are purchased from wireless vendors such as Singular, Sprint, and Optus MobileMail, so that limits their availability somewhat. One neat item we noticed was the ability to remotely remove data from a lost device. This is certainly a nifty option and offers that little bit of extra comfort should your PDA or phone be lost or stolen.

Accessing mail directly from Exchange using POP is not really recommended unless you have the added capability installed to enforce SSL encryption over the link. As we mention earlier, POP passwords are in clear text as they cross the network so can be intercepted.

Connecting directly to Exchange

Okay, so you decide that you do not want to use POP and want more complete Exchange access. There are numerous methods for this, including some of the tools we mention in the previous section. Service providers often offer this capability for additional cost, as well. Using Rogers in Canada, Barry set up his Treo 600 to access an Exchange server. You see in Figure 8-1 that it requires a desktop redirector to accomplish the connection.

Figure 8-1:
Using a desktop redirector to access Exchange.

What happens is this. You set up an account with your service provider and then download the necessary software for both your workstation and your PDA. After installing them on their respective devices, you configure the system, using the vendor's Web interface. The redirector allows your PDA to access your e-mail using the workstation you set it up on, so that machine needs to always be connected to your LAN and the Exchange server. You can see the Web page that indicates your service is ready in Figure 8-2. Note that this service differs in implementation among vendors, so you need to read the documentation that arrives with your product.

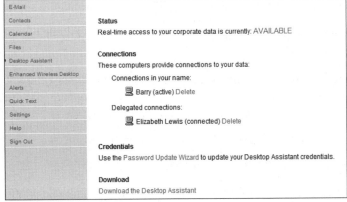

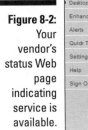

Figure 8-2:
Your
vendor's
status Web
page
indicating
service is
available.

Obtaining Exchange or Lotus Notes e-mail requires that you understand the logon process for your mail system and its address on your network. Barry's service provider lists the following requirements for corporate e-mail access:

- Microsoft Outlook 97, 98, 2000, or XP
- Lotus Notes 5.0
- Exchange Server 5.5 or 2000
- Lotus Domino 4.6 or 5.0
- A Windows workstation (95, 98, NT, 2000, XP)

Outlook Mobile Access and Exchange

Perhaps you have used Microsoft's Mobile Information Server in the past to allow for mobile devices (such as PDAs) to interface with Exchange. Microsoft Server 2003 incorporated these changes and now includes those functions. Those of you using this newest of Microsoft's operating systems have direct access to wireless e-mail.

Outlook Mobile Access provides browser access using your mobile phone and XHTML (WAP 2.0), or an HTML mobile phone and PDA browsers (such as Pocket Internet Explorer) offered on Windows Powered Mobile Devices. Using this access, you get full support of e-mail folders, calendar, contacts, tasks, and even searching the Global Address List or personal contacts. It's like having workstation access to the Exchange server.

Although Mobile access is built into Exchange, it is not enabled by default. (Microsoft gets some things right, after all.) To enable it, start Exchange System Manager, double-click Global Settings, and then use the Mobile Services Properties dialog box, selecting the options as needed under the heading Outlook Mobile Access.

Setting up the system entirely is beyond the scope of this book, but the key steps are:

1. Configure your Exchange 2003 front-end server for Outlook Mobile Access.

2. Enable Outlook Mobile Access on the Exchange server.

3. Configure user devices to use a mobile connection.

4. Instruct your users in using Outlook Mobile Access.

Armed with this information, some time, and a little (okay, a *lot* of) patience, you can get yourself wirelessly connected to your Exchange e-mail.

Outlook Web Access and Exchange

Perhaps you have used a Web browser to obtain e-mail from Microsoft Hotmail or are planning to use Google's new Web-based e-mail service. If so, you might also know that you can access your Exchange server by using a browser and your wireless access device.

Outlook Web Access has been around for some time and appears to be mostly used within the college crowd, although there is some business use. By default, Outlook Web Access is enabled in Exchange 2003. You may perform the following general steps to complete the installation and allow your users access:

1. **Set up a logon page.**

2. **Configure authentication.**

3. **Configure security options.**

4. **Configure Outlook Web Access compression.**

You can set up a logon page for Outlook Web Access that stores the user's name and password in a cookie instead of in the browser. When a user closes their browser, the cookie is cleared. The cookie will also be cleared automatically after a period of inactivity. This page will require the user to enter their credentials to access his e-mail.

We strongly recommend that you consider using a third party two-factor authentication product such as Securid instead because this provides far stronger authentication and helps eliminates some potential issues with denial of service attacks. Web-based authentication leaves your Exchange server open to brute force password attacks from the Internet. Using easily obtained tools, unauthorized persons can run automated logons against your network possibly gaining access to accounts through their use of weak passwords or company defaults.

Now you need to enable forms based authentication in Exchange. You do this by setting the Enable Forms Based Authentication option in the Outlook Web Access Settings dialog box. Make sure that the time-out parameters are set to disable the session within a reasonable timeframe of inactivity, such as 15 minutes (the default setting). This helps prevent unauthorized access if the user forget and leave his session running while wandering off for a coffee break.

The default Outlook Web Access logon page enables the user to select the security option that best fits their requirements. It uses two settings, Public or Shared Computer and Private Computer. The Public or Shared Computer option is selected by default and provides a default time-out option of 15 minutes. The Private Computer option allows a default time period of 24 hours. Essentially, this option is intended for users who are using personal computers in their office or home. We suggest that the options be set within your login page to 15 minutes and no option be provided for the user to change the time period. The small aggravation in needing to authenticate after 15 minutes is easily outweighed by the potential for loss if a user chooses the longer time-out period on a public computer and then forgets, leaving their session open to all.

Finally, compression is available to enhance slow network connection. This is especially useful for your wireless access users. Three settings are available depending on whether your Web site uses static, active, or both types of Web pages. Compression depends upon Exchange 2003 running on the Windows 2003 platform with the user's mailboxes stored on those machines. It doesn't function with mailboxes stored on legacy Exchange 2000 servers.

These basic steps guide you through the rudiments of setting up Microsoft Outlook Web Access. Be sure to read your Exchange documentation and visit the Microsoft Web site to obtain truly detailed information before venturing down this road.

There are a number of considerations when thinking about using Web-based access to your e-mail. These range from weak passwords, possible lack of virus prevention, and user data remaining on workstations.

Allowing user authentication directly to your Exchange server poses a fairly major risk of unauthorized access. In our experience, users choose poor passwords, and these can therefore be easily attacked. A number of tools automate logons, allowing a hacker to try thousands of logins within minutes. The likelihood of finding those users with weak passwords is almost certain. In addition, merely attempting to login numerous times to each account will invoke the lockout parameters that your security department has set, effectively disabling those accounts and preventing legitimate logins.

If your users plan to access their e-mail while traveling and use a public computer, they might inadvertently attach a virus to their e-mail, affecting your inside network unless you run an antivirus product on the Exchange server or firewall. We know of corporate clients who have yet to install antivirus software on their Exchange servers, citing difficulty in doing so but thereby leaving these machines vulnerable to a virus attack. A strong antivirus implementation is a necessity.

Finally, using Web-based e-mail leaves any file attachments you might have in temporary folders on the workstation you are using. Someone can obtain these after you leave, exposing your corporate secrets to unauthorized access.

Outlook Web access offers a neat method for getting your e-mail but is not without its risks. Consider your options carefully before implementing it.

Wireless Hot Spots: What's New Around the World?

Wireless is changing almost overnight around the world. Hotels, airports, cafes, and restaurants are adding hot spots every day. All these work to enable you to remain connected to your office, possibly allowing you to resolve those technical issues while sitting in a hotel or airport lounge. Finding the currently available hot spots is the key. You might try using one of the many Web searches to do this before you travel to that new city on business or pleasure. One site, `www.wifinder.com`, allows you to search for both public and private hot spots around the world. Other locations include using your commercial dial-up vendor if they have evolved to include the wireless world. We use AT&T Global for obtaining dial-up around the world. So far, they remain committed to offering only dial access. In Canada, however, Allstream (`www.allstream.com`) offers not only dial connectivity but also wireless hot spot roaming, extending your ability to remain connected.

In order to keep up with all the changes, you need to keep a close eye on what's happening if you plan to be connected any time soon. We are starting to notice new uses of wireless access, such as Voice over IP (VoIP), which will begin to change the way we connect with one another, possibly reducing the use of Mr. Bell's original invention and relegating it to the bone-yard. Imagine wanting to make a telephone call and using your laptop rather than a cell phone merely because you are already logged in somewhere and it's more convenient — and thinking little of it!

An enterprising Web site at `www.guerrilla.net/freenets.html` provides a list of wireless hot spots around the world. Look for a number of such sites to spring up as services expand and the user communities respond.

In the air

An interesting new development is in the air — literally! Recent announcements indicate that soon you may be surfing the Web and connecting to your corporate e-mail while flying high, 38,000 feet in the air. A company called Connexion by Boeing is beginning a foray into the wireless world with a difference. They are not targeting buildings; they are targeting airplanes.

The service offers connection via wired or 802.11b wireless connectivity. Lufthansa began offering the service in May 2004 on flights between Europe and the United States. Rival Tenzing offers a scaled-back version that permits e-mail access stating that its research found that most passengers (around 86 percent) want e-mail access for the most part and are less interested in browsing the Web while high in the clouds. Tenzing service is available on some Cathay Pacific and Virgin aircraft among others.

With most new laptops capable of wireless connectivity, airlines may find yet another compelling reason for wireless over wired access: less weight. With no need to install cabling throughout the plane, there is a small gain to be found. When every ounce counts in terms of high priced jet fuel, the advent of wireless makes more sense.

Wireless connectivity is managed in different ways by the vendors. Connexion accomplishes this by installing an access point on the plane that interacts with satellites high above to provide near seamless connectivity even while traveling at a few hundred miles an hour. Rival Tenzing uses a store-and-forward server that forwards the e-mail and as a result does not allow VPN access. These solutions offer ways to ensure you remain connected to your office as you fly across the country, using your travel time to become more productive.

New ideas for wireless network attacks

One interesting item we noticed recently concerns a small airplane developed by AeroVironment called the Wasp Micro Air Vehicle. This little pint-sized plane has a wingspan of 13 inches and can stay aloft for about 2 hours. Apparently DARPA, the US Defense Department's research arm, is looking at it for battlefield reconnaissance using small cameras. So what has this to do with wireless, you ask? Well, imagine if some competitor wanted to use a wireless PDA with automated data sniffing software installed. They might have two hours to hover within range of your wireless network with no one the wiser. Far-fetched? Perhaps, but it may only be a matter of time before this level of attack occurs.

Expect to see a lot more identity theft in the coming years. As more home users migrate to wireless, they leave their computers possibly even more vulnerable to attack than they did previously with wired connections to the Internet. It isn't hard to imagine nefarious persons roaming around huge apartment complexes scanning for wireless networks and then trying to attack them. With the home address already predominantly identified by the physical location, scanning e-mails, file transfers, and any other home traffic, hackers will get access to all kinds of useful data that can be parlayed into identity theft.

Part III
Using Your Network Securely

The 5th Wave   By Rich Tennant

"We take network security very seriously here."

In this part . . .

In this part, you discover how to protect the investment you've made and the data crossing your wireless network. You find out all about the risks to your network, clients, and data, and you see how to design a secure wireless environment to protect against those risks. Designing and deploying a *secure* network is probably the last thing you want to think about as your network becomes available and you want to use it, but we caution you against skipping this part. If you skip it, you'll quickly regret it when your data is stolen or your network is used by unauthorized persons.

This important part shows you all about using good security techniques, including the basics of WEP and WPA and moving into advanced security with EAP protocols and AES encryption. Finally, you see how using VPN technologies can be a boon to securely accessing your network and keeping the bad guys out.

Chapter 9

Considering a Deadbolt: Understanding the Risks of Wireless Networks

*W*ireless networks are wonderfully freeing devices, allowing you to roam from your desk while using your network. In fact, you can connect while traveling around the world — that has to be a really neat thing, right? Now it is time to discover the perils of all that freely accessible access. In this chapter, we show you how being too cavalier with a wireless network can cost you in terms of time, money, and loss of business information — possibly to your competitors.

Risks to the Network

A network is always at risk. Whether wired or wireless, there are many ways that unauthorized access can occur. In your wired network, if you allow casual physical access to your business premises, someone you do not know can attach to your network and start stealing information. A simple example is letting an unknown salesperson use an empty conference room without supervision. Most businesses enable these rooms with network access, so it is simply a matter of plugging into the wall socket and starting some hacking tools. Barry has used this very method during client engagements and obtained enormous amounts of data about the client network, including obtaining sensitive data, prior to any help from the company. Of course, he did this after obtaining their permission to do a network penetration exercise. In one memorable assignment in the hills of Boise, he and a colleague spent a few days in a conference room, only appearing for lunch and to go home, without talking to anyone in the client site. Eventually, he set up a meeting and showed management the results, which included user accounts and passwords, their business plans for the coming year, and more.

A wireless network is even easier to access. You see later in this chapter that there are groups who have nothing better to do than go around the country locating and marking companies that use wireless networks. They even physically mark the location so others walking past can see. You recall from Chapter 2 that your wireless access point broadcasts itself for some distance, depending on the version. That typically extends beyond the boundaries of your office walls.

Coupled with this risk is the potential for jamming your transmission or gaining access through your use of default passwords. It's a rough, tough world, and you need to learn the issues and how they might impact you.

Going to war: War nibbling, war driving, war flying, and war chalking

No, we don't mean war with guns and tanks. This is information warfare — discovering wireless networks and then sometimes using or attacking them. When you broadcast your wireless access point past your building's boundaries, you are bound (pardon the pun) to attract attention, and unfortunately, that attention includes things like war driving and war chalking.

These methods of war arrived with the advent of the wireless local area network (LAN). They follow the basic premise of attempting to find access points and show where they are to others. It's become a game, albeit not a nice one, among many people. There are numerous Web sites dedicated to this topic, including www.geekzone.co.nz and www.seattlewireless.net/index.cgi/WarDrivingSoftware.

War nibbling

War nibbling is similar to war driving, but it's only against wireless personal area networks (WPANs) and the Bluetooth technology. War nibbling involves locating and identifying wireless connectivity and the inherent security in place (or not in place). There is a good article about war nibbling on the @Stake Web site (`www.atstake.com/research/reports/acrobat/atstake_war_nibbling.pdf`) that provides you an idea of how this works. You recall that Bluetooth technology typically operates at smaller distances, and that means you need to be closer to detect it. Sorry, no sitting in the park on a sunny day (unless folks are using Bluetooth around you). More devices than ever incorporate Bluetooth, though, so look out for those laptops, PDAs, and cell phones while you prepare for war nibbling. So how do you locate Bluetooth devices? Well, one way is to look for PDAs and laptops with your trusty little eyes. But that isn't really effective, is it? Not all of these devices are Bluetooth-enabled. In fact, none of my many Toshiba laptops is Bluetooth enabled. Many vendors make them Bluetooth capable, but require additional cost add-ons to enable it, which many people don't bother purchasing.

A better method for finding Bluetooth-enabled devices is to download the tool called Redfang from the @Stake folks (`www.@stake.com/research/tools/info_gathering`), install it on your Linux laptop, and then go hunting. This advanced tool allows you to find Bluetooth devices that are set to non-discovery, a technique that was designed to try and protect devices when their users did not want to share with others. Fortunately, new Bluetooth devices with version 1.2 are not prone to this attack. Whew! Guess I'll check the version of the next device I purchase.

War driving

War driving is already the granddaddy of the war line. Okay, it's a young granddaddy — the wireless community isn't that old. It became immensely popular after the advent of wireless LANs and involves finding all those 802.11a, b, or g access points you've installed. Barry has taught a number of network penetration seminars around the world where he demonstrates the ease of finding vulnerable access points. One of the few places he had difficulty was in Kuwait last year, but wireless access is only beginning to intrude on that market. He once showed a class in Melbourne, Australia, how many access points were available right around the hotel (quite a few as he recalls), and few of them were secure.

So how is this accomplished? Glad you asked. First, if you are unsure of the popularity of wireless access points in North America, visit `www.netstumbler.com/nation.php` and look at the map provided. If you plan to drive across the country and war drive along the way, you'll note it's best to stick to the west coast and east coast if you really want to locate devices. There's not a lot going on in North Dakota or New Mexico.

In order to locate wireless access points around the country or around your neighborhood, you need a toolkit. This consists of the following:

- ✔ A laptop computer
- ✔ A wireless network card (although you may have AirPort or a Centrino chip)
- ✔ An antenna
- ✔ A car (okay, I guess you could use a bicycle, but it limits your range)
- ✔ Software for locating access points

The first point is fairly self-explanatory. Any recent laptop will do, although you might want a later version of Windows running because device drivers might be harder to get if you are still stuck on Windows 98 or 95. You can also run a Mac or your favorite version of UNIX.

You need to be aware that there are some restrictions on the network cards you can use to do this type of work. NetStumbler lists the following cards as working with version 4 of the software:

- ✔ The Proxim models 8410-WD and 8420-WD. The 8410-WD has also been sold as the Dell TrueMobile 1150, Compaq WL110, and Avaya Wireless 802.11b PC Card.
- ✔ Most cards based on the Intersil Prism/Prism2 chipset.
- ✔ Most 802.11a, 802.11b, and 802.11g wireless LAN adapters on Windows XP machines, although NetStumbler indicates that some of these may also work on Windows 2000. The Windows 2000 implementations may report inaccurate signal strength, and, if using the NDIS 5.1 card access method, the noise level will not be reported. This includes cards based on Atheros, Atmel, Broadcom, Cisco, and Centrino chipsets.

We have used Proxim (Orinoco), Alvarion, and SMC cards in an Intel laptop with great success.

Using an antenna is optional, but it greatly increases your ability to identify and find wireless networks. We purchased external high-gain antennae from Hugh Pepper (mywebpages.comcast.net/hughpep). These greatly increased our range. If you don't want to spend additional funds, however, the antenna in the wireless access card will provide you with numerous wireless locations as you drive around your town or city.

War driving naturally infers that you are driving. You can do this by merely walking around at lunchtime. Driving only adds the ability to cover more distance in less time and therefore discover more locations faster. Take your bicycle out for a spin and balance your laptop on the handlebars. You'll look a little weird, perhaps, but you can still obtain access to those wireless sites you ride past. Another item of interest is in interacting with others who may be doing the same thing at the same time. Using an application called Automatic Position Reporting System (APRS) allows you to display the location of fellow war drivers, which gives you a chance to communicate with them using a chat program or even two-way wireless radios. There's power in numbers. You can find the program at `www.cave.org/aprs`.

Lastly, you need software, and the most well known is NetStumbler. Find it at `www.netstumbler.com`, where you also find great information on the latest and greatest in this arcane department. A version called MiniStumbler exists for those with Windows CE PDAs. Mac and UNIX users can look in Chapter 17, where we list versions for those platforms.

With all these components available to you, you can put together one really awesome toolkit. Install the wireless access card on your system first, and then install the software. On most of the operating systems, this is fairly simple to accomplish. After installation, run your software and ensure that it is locating wireless devices by seeing if it finds the access point you installed in Chapter 5. Make sure that your wireless access point is turned on and running, of course. You should see something like what is shown in Figure 9-1.

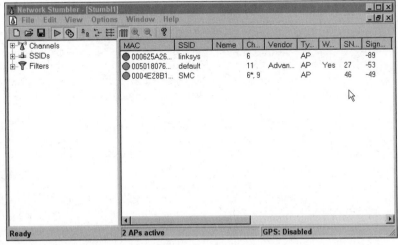

Figure 9-1:
Viewing
wireless
networks
using
NetStumbler.

NetStumbler automatically begins by showing you networks within range of its associated wireless access card. This is because the option Enable Scan is selected under the File menu. If you want to start and stop it manually, uncheck this option. The File menu provides you with methods to save the logs that you capture for later review using the Save As function. Note that under the View⇨Options menu you have the capability to add a Global Positioning System (GPS) receiver to NetStumbler. You need a GPS that supports NEMA format, which now supports serial and USB connectors. This allows you to log the actual latitude and longitude coordinates with mapping software like Microsoft's Streets & Trips or MapPoint to produce a detailed map of all the access points you find. It makes for a great report or provides for general Internet use, as you see if you go to www.nakedwireless.ca/winudcol.htm.

After all the parts are put together and you are assured it is working by viewing your own access point on the software, then happy hunting!

We must add one caveat, of course. We do not condone illegal activity. Finding wireless sites is one thing; trying to use those sites for nefarious purposes (which include any access at all) is not only wrong, but it may be illegal, depending on where you are located.

There are recorded cases in which law enforcement has charged people who have accidentally associated with an access point. Windows XP users are particularly exposed. We recommend that, when you go war driving, you unbind the TCP/IP protocol from your wireless adapter. Better safe than imprisoned.

War flying

The bad news is that war driving includes flying airplanes to find wireless networks. The good news is that it is less of a risk to your wireless network because the person flying needs to stay motionless to obtain any reasonable number of data packets. So, if you see a stationary helicopter hovering around your house or business, you may want to make sure your network is secure.

It appears the first people to exploit this form of finding access points were some people in Australia, although the folks from California in the site mentioned below apparently published results first.

According to the site arstechnica.com/wankerdesk/3q02/warflying-1.html, war driving is passé, and war flying is in vogue. Brian Grimm, spokesman for the Mountain View, California, based Wi-Fi Alliance, mentions that an altitude of 2,500 feet appears to be the limit for wireless access from the air. I am not so sure about the potential risks myself, as I believe the risks are far greater with land-based exploits because of the potential for stationary data collecting. Regardless, as we extol in our security chapters, make sure you are encrypting and properly securing your wireless connections.

War chalking

It's bad enough to know that perhaps your wireless network makes it into one of the sites we mention and to know that the world knows about you. Consider those who are subjected to war chalking. *War chalking* is the physical marking of your site with special symbols, sort of like graffiti for wireless network weenies.

There are many Web sites that provide details on war chalking, one of which is www.warchalking.org. Here you find the details of what symbols are used and how those symbols direct individuals to information about your wireless access point. It is apparently inspired by the Depression-era practice of hobos marking homes that were friendly to them. The three main symbols revolve around an Open node, a WEP node, and a Closed node. Figure 9-2 depicts the symbols.

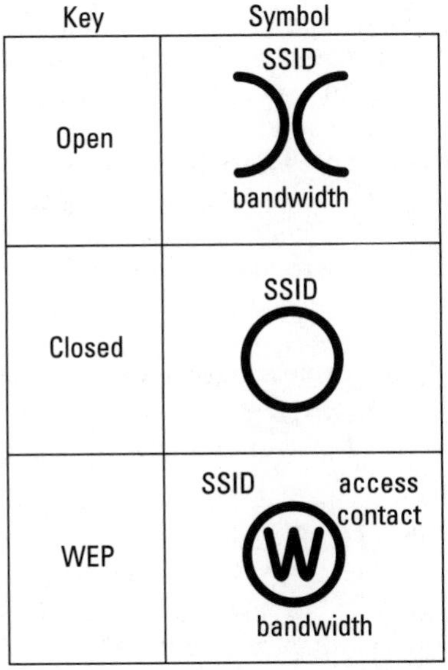

Key	Symbol
Open	SSID)(bandwidth
Closed	SSID ◯
WEP	SSID access contact (W) bandwidth

Figure 9-2: War chalking symbols.

Have a look around next time you wonder the streets of your city and see if you can locate these symbols. It allows you to emulate war driving without a laptop, wireless card, or software. Send us an e-mail if you find any symbols (because we think it is an urban myth).

A roguish WLAN

In movies, the rogue is often debonair and dashing, as in *The Rogues of Sherwood Forest*, a 1950s film about Robin Hood. "Steal from the rich and give to the poor." Great entertainment, but not very nice when you are the rich with your wireless network and someone decides to give to the poor by allowing others access to that wireless forest of yours.

Adding fake access points to your network is one way someone can increase the spread of your network and provide ready access to people farther away than you thought. This can be done by a staff member who, for example, wishes to sit in the park and connect so you think she is still working. Well, she may still be working, but so are all the hackers lurking around this newly discovered access point.

Using the information discussed in the earlier section, "War driving," an attacker can readily configure a rogue access point using your SSID, WEP keys, and MAC addresses. Argh! This can enable them to create a typical man-in-the-middle attack by adding a rogue access point, getting you to use it, and then intercepting all the traffic you send through the access point. Using a rogue AP, an attacker gains valuable information, such as authentication requests, the secret key that is in use, and, of course, any data that you may transmit. To avoid detection, the attacker sets up his machine with two wireless adapters: One card is used by the rogue AP, and the other is used to forward requests through a wireless bridge to the legitimate AP.

You also need to be aware of how your wireless network is susceptible to other types of attacks and what to do about it.

Open broadcast of SSIDs

You recall that the service set identifier (SSID) is used as a rallying point to differentiate one network from another. It basically acts as a clear text item that can be seen by all those war driving or war flying past your business. Anyone who needs to connect to the network must first enter this SSID in her wireless utility. We show you how this works in Chapter 6.

Openly broadcasting your SSID makes access to your network a trivial matter for people in the know who enter it on their own network cards and thus gain access to your wireless network, depending on what other security you implement. Revisit Figure 9-1 and see that the SSID is clearly visible in NetStumbler.

The good news is that you can disable this broadcast feature on most access points. The bad news is that if you are using older models, like my old SMC Barricade 7004AWBR, you have no option for disabling the broadcast. Check your manual to see how this is accomplished on your particular gear. The worse news is that you may not want to disable the SSID.

What's that? Don't disable it? Remember, the SSID is not designed to be a security tool, and disabling it may have adverse effects on your network. If you have a really small network with only one or two access points, disabling the SSID will not likely cause you harm. However, in a larger network with multiple access points and mixed client deployments, it may be more trouble than any supposed benefit. Cisco advises leaving the SSID in broadcast mode to prevent any problems. Be aware that disabling the broadcast mode really doesn't gain you anything because the SSID is still visible within the probe response frames. It merely means that the SSID cannot be seen by using NetStumbler; but if you use Ethereal, Commview, or some other more capable packet sniffer, you will see the SSID.

Bottom line for disabling your SSID: It might buy you a small peace of mind, but if that's all you rely on for security, you will quickly be awfully surprised and upset. (Check for chalk marks around your building.)

Jamming

Jamming relates to someone taking your device off the air by overriding your wireless access point's signal with a stronger signal. This occurs both maliciously and by accident. In the 2.4 GHz range, your cordless telephone can interfere with your access point signal. Barry once had a phone that blew away his wireless connection every time he answered it. It was darn annoying — enough so that he got rid of the phone. The 802.11b band is particularly susceptible to such interference; therefore, eliminating it by using 802.11a, for example, is often a solution, albeit a more expensive one.

Wireless jamming in the cell-phone environment is fast becoming a divisive issue, with proponents arguing that jamming signals in theaters, for instance, allows for a peaceful experience without a cell phone ringing in the middle of the act. On the other side are those who believe it's an infringement of their rights and isn't to be allowed. When dealing with our wireless access points, the issue is easier to define: We need to ensure that our signal reaches our users.

Using jamming equipment, our competitors can put our business at risk, especially if our business depends on wireless access. Consider a hotel offering wireless access only to have guests constantly complain they cannot get a signal. There are many jamming methods available, including using professional jamming gear. Sites such as www.globalgadgetuk.com sell various tools offering, for example, cell-phone jammers. One really neat tool they offer is a handheld unit that a person can use anywhere. Imagine going to lunch or a movie and disabling all cell phones on the 800 and 1900 MHz bands. In Europe, you'd use their 900 and 1800 MHz version.

Jamming access points requires Global Gadget's model 2.4JM (www.global gadgetuk.com/wireless.htm), a handheld unit that also jams Bluetooth connections. Its effective range is said to be about 10 meters. Imagine the fun you might have going into the office and turning it on and off during the day. Not that we condone such activity, of course, and jammers are illegal in some places, so be aware of the laws in your part of the world. Consider all those intermittent errors you may have in your network and how they may or may not be caused by jammers, and then you'll realize just how tough problem solving can be in this area.

On the plus side, jamming can also be used to eliminate rogue access points on your network. A product called AirMagnet Distributed 4.0 provides a tool for locating rogue access points using a combination of techniques, such as comparing the MAC address, SSID, and manufacturer, to determine who is permitted on your network and shutting down others by blocking them from your network. You can see more details at www.airmagnet.com/, including the numerous operational tools they offer to help manage and secure your wireless network.

Other vendors offering similar solutions include Enterasys (enterasys.com/ home.html) and AirDefense (www.airdefense.net). These companies have great product lines that will enhance your security immeasurably. Okay, no plugging firms. Honestly, we have received no payment from any of these firms and only offer them in the sincere belief that they can improve the security of your wireless network.

Signal loss

Losing your signal is next on the list of bad things that happen to wireless network broadcasts. Signal loss occurs in many ways, starting with the normal loss that occurs in the cables and devices you use, as we point out in Chapter 2. Of course, it also occurs through the jamming we mention in the preceding section.

The first thing to realize is that loss occurs naturally. As your wireless signal propagates through the air, the laws of physics intersect, and the air itself eventually becomes a factor in limiting how far your signal travels. This loss, or *attenuation,* as the industry refers to it, may or may not be a problem for you. In a small wireless network within the walls of an SMB, you may never experience signal loss. Your signal is strong enough to accomplish your needs, and while it does lose its strength eventually, that is far outside the parameters of your connected workers. So don't worry about it.

In larger firms, or those needing greater distance, signal strength needs analyzing and resolving. In Chapter 2, you see how to account for the signal you need and what actions to take to ensure you get it. Now you need to verify and ensure that you are obtaining the needed results.

Remember that all objects cause some form of signal loss or attenuation, and these are part of your earlier calculations. However, perhaps you have changed your physical premises, moved a few walls, or built a conference room on the main office floor and now are experiencing loss. Redo your calculations or build in a few general loss figures into your original plan to account for the changes. You may need larger antennae or more access points. Table 9-1 shows some of the loss rates for general objects.

Table 9-1	Signal Loss from Common Objects
Object	*Loss*
Plasterboard (gyproc) walls	3 dB
Cinder block	4 dB
Glass wall with metal frame	6 dB

Vendors may boast that their gear has a range of 300 feet. This is obviously under ideal conditions. As a rule of thumb, you lose about 20 feet when traveling through an interior wall. When you consider that your signal may travel though numerous layers of cinder or plasterboard, it may be little wonder that the office five walls away is having difficulty getting a good signal. Incidentally, people make good barriers also. Although we could find no actual statistics, anecdotal evidence points to the idea that a roomful of people can prevent wireless access from an access point whose signal is already weak.

Risks to Your Users

Along with network risks, you must consider your users and the risks they face. This may be more of an issue than the actual network risk you face. In fact, it probably is more important. Is it critical to figure out which is more serious a loss? Not really. They are interrelated enough that you need to take care of the network and the users equally and provide for a sound degree of security over both.

Your users face different risks, however. From people who try to steal their very identities, to the typical weakness of poorly designed and default passwords, you must keep an eye on how the users use your network, and how they may leave it open to attack.

Target profiling

The first piece of business for anyone trying to attack your network involves *profiling,* or *fingerprinting.* This means finding out who you are, whether your network is worthy of connection (if the intent is to perform corporate espionage), what device brands you use, your SSID, WEP keys, number of visible access points, and any other data that may be useful.

You may be amazed by how much information can be garnered this way and how open that leaves your business to unauthorized access. If the profiler is really gutsy, she might even attempt to get information through social engineering. One aspect of social engineering is pretending to belong to the company and asking the help desk or other users to provide you with information in the belief that you are permitted access to such information. If you are an open and honest person, this takes advantage of that attribute and abuses your trust for personal or professional gain. Most network penetration assignments include aspects of social engineering because it is the easiest method to get information.

Identity theft

Stealing your identity sounds like science fiction but is unfortunately more common than ever in this electronic age. Adding a wireless network exacerbates the problem because many organizations do a poor job securing the network.

What is identity theft? In a nutshell, it is someone pretending to be you, using your name, age, address, social security number, and even credit card data. This allows the identity thief to enter into legal agreements (to buy cars or houses) or get loans or credit cards approved — all in your name. Obviously, these nefarious folks then take off and leave you with the bill. The Federal Trade Commission states: "Identity theft occurs when someone uses your personal information, such as your name, Social Security number, credit card number, or other identifying information, without your permission to commit fraud or other crimes." So protect that information!

This is even more crucial for those of you who use this technology at home. At home you are less likely to implement the very items that protect you, either due to carelessness or in the belief no one will notice and target your home network. After all, they are your neighbors, right? If you are a business that provides this technology for your staff to use at home, insist on a strict security regimen and have staff attend training before they take equipment home.

Any ability to access a network with ease allows an unauthorized person hours or even days to figure out any internal security you may use and to obtain access to the personal details kept on your computer. Think using that Microsoft Money or Word password is going to deter them? Not a chance. You can buy software to bypass hundreds of different password-protection schemes on the Internet. The only true protection is not allowing anyone in by securing the network and, if necessary, turning off your computer whenever it is not needed. Obviously, not keeping personal records on a computer goes a long way, but who does that nowadays? You wouldn't be reading this book if you didn't use computers extensively.

Lack of authentication

The need to ensure we are who we say we are is fundamental to a good security program. This is where authentication enters. Unfortunately, there are numerous methods to perform this function, and some are better than others. Effective authentication techniques take time, effort, and training, and are therefore sometimes omitted.

Default postures on most access points offer no authentication of your users and are therefore a problem, unless you plan free access to the outside world.

If you want to be sure that only your authorized users access your wireless network, you need to consider advanced techniques such as EAP, LEAP, and PEAP. Chapter 11 discusses these techniques in detail.

Relying on WEP and MAC filtering for security does not provide you with good security. In fact, it can be argued that it provides a false sense of security because you believe that security is in place and therefore put more data at risk than you might if no security was used at all. A protocol that authenticates users with WEP keys and MAC addresses is easily defeated and will ultimately lead to your site being compromised — it's only a matter of time. That being said, however, we strongly urge you to use at least that level of authentication because, to paraphrase a well-known saying, it is better to have done one thing than to never have done anything. Conflicting statements? Perhaps, but we know that many will ignore our warnings. We see it all the time. So even though WEP isn't the best protocol to use, it is better than nothing. Perhaps attackers will take the easy way and attack your neighbor's completely unprotected network instead of yours.

Make sure that you train users on the importance of using well-designed passwords and to not keep them on their laptop in a file called My Secret Passwords. Organizations lose laptops in alarming numbers, and besides the corporate data that is potentially lost, providing an attacker with your authentication credentials is just plain foolish. Don't forget those PDAs, either. They are becoming popular and are being used for access via VPNs and other methods by technical support staff. This means those devices usually have a lot of access to your system, and, if lost with either a password file (or worse, with automatic log-in capabilities), they pose a serious threat. Users need to be made aware that passwords are critical and are never to be written down, saved on the hard drive, or automatically saved by log-in procedures.

Make sure that you read and follow our directions for security in the later chapters. And remember: It is unlikely that what you have is unimportant, so keep it safe.

Default passwords are de fault

Ah, default passwords — a wonderful invention. Made by vendors to ease their support calls when customers install their software. Anything to make their lives easier. No point thinking about security that early in the game, right? After all, they do tell you to change the default!

This is one of the most insidious exposures in products today. Reading the latest news, we see that Cisco recently admitted that a default hard-coded account exists with a known, fixed username and password combination in some versions of their Cisco Wireless LAN Solution Engine (WLSE). The WLSE provides centralized management for Cisco Wireless LAN infrastructures, leaving your Cisco wireless network vulnerable until you apply their patch.

A hard-coded password. In 2004. Astonishing. We have been against organizations hard coding passwords for over 20 years now, yet we still have vendors not taking security seriously enough.

Regardless of the Cisco debacle, and they are unfortunately not alone in this, ensure that all the defaults you are aware of in your access points, and everywhere else for that matter, are changed to something strong. We mention earlier that all access points arrive with a default password, sometimes even a blank password. This leaves it possible for anyone to log in and change all your security options unless your changes include a strong password. Ensuring that all your network components, such as routers and switches, as well as your operating systems have strong security enhances your chances should you be attacked. Play it smart and realize that security is just a way of life for you now.

Risks to Your Data

Keeping data secure and safe from unauthorized access is the *raison d'être* for an entire security industry. Thank goodness, or we would be out of business. That said, however, the risks increase in the wireless world. As you've already seen, default passwords, lack of security, and many other reasons leave numerous wireless implementations sorely lacking and vulnerable to attack. In Chapters 10 through 12, we show you how to mitigate these risks. On a positive note, a recent "Report on Technical Standards" released by a CyberSummit taskforce on security in the United States made some very promising recommendations for vendors to follow in their product life-cycle. There are over 20 recommendations including the following:

- ✓ Produce more realistic security testing of products using real-world situations.
- ✓ Provide better security recommendations, configuration checklists, and best practices in product documentation.
- ✓ Make products secure by default.
- ✓ Include a tool or capability that allows a user to quickly and easily report on the security posture of the installed product.

These and all the other recommendations, if followed, will lead to a more secure environment and will require less effort on behalf of system staff to ensure a sound overall security posture. You can find the report at www.cyberpartnership.org/TF4TechReport.pdf, the Web site for the group that produced the report.

You call that encryption?!

The more common encryption protocol for access points is Wired Equivalent Privacy (WEP). But before we get into that, a really brief primer on encryption is necessary. In a nutshell, encryption is the process of turning a cleartext message into a data stream that looks like a random sequence of bits, hiding the actual clear text message. How this is accomplished is way beyond the scope of this book, but if you really need to know, purchase *Cryptography For Dummies* by Chey Cobb (Wiley).

So you want to hide your cleartext from others yet allow those you want to see the original message. WEP performs this step in your basic access points. When implemented, each time a user connects to the access point, his network packets are encrypted across the wireless airwaves and are decrypted by the access point. This means that encryption is only useful on the wireless portion, and, after you connect to your wired LAN, the data is no longer encrypted. This is usually fine because you are attempting to protect the more vulnerable wireless network.

WEP uses two key lengths. This is where the base strength of the encryption is derived. It's like having a really locked down server: It's very secure unless you happen to have a weak administrator password. The key can be likened to the password. Your secret key is typically a 40-bit number or a 104-bit number. This is increased by WEP through a 24-bit initialization vector (IV) number that is managed by the software. You often see vendors touting a 64-bit key and a 128-bit key. 64-bit WEP is the same as 40-bit WEP! The lowest level of WEP uses a 40-bit user key with the 24-bit IV. It's just that some vendors refer to this level of WEP as 40-bit, others as 64-bit.

So WEP then uses the shared secret key you supply and the 24-bit initialization vector as the complete key. It is the use of this random IV and a static user key that weakens WEP security. Most people rarely change their WEP key. This, combined with the small initialization vector, allows a persistent hacker to eventually crack the key and access all your encrypted data. We provide you with some of the tools to test this for yourself in Chapter 17.

Some vendors are addressing this weakness with larger keys, such as Agere Systems with a 152-bit key and D-Link with its 256-bit key length, but these are also susceptible to attack; they just take longer to crack because they are not addressing the inherent WEP weakness. The new 802.11i protocol looks to address this fundamental weakness. Of course, you can always implement a VPN solution, which would dramatically improve your overall security, as we show you in Chapter 12.

Accidental associations

Your wireless network usually cannot be easily contained within your organization; therefore, accidental associations can occur with neighboring networks.

The WLAN-friendly Windows XP operating system in particular makes it easy to enable your wireless users to automatically associate and connect to this neighboring wireless network without your users being aware of what is happening. To know whether you have this problem, you can visit `www.wigle.net`, an active site that collects wireless access point locations, over a million locations listed. It might be illuminating to see all your neighbors listed. If you enter Boston in the city search section, for example, you see a massive map covered in red, indicating wireless networks.

Whether you're talking guilt or network connectivity by association, you need to be aware that you might connect to the wrong network without realizing, and therefore send confidential data across someone else's network. In fact, it's not hard to imagine installing one on purpose in the office next door in order to try and steal your trade secrets. The ultimate defense against this type of attack is to purchase defensive hardware such as that from AirDefense or other vendors that we mention earlier in this chapter.

Eavesdropping

It isn't difficult to eavesdrop on wireless connections, even if it may be illegal or at least unethical. In the wireless telephone industry, as with your wireless network, you basically use radio transceivers to accomplish your call. Your voice or data transmits through the air on radio waves. You receive the data from the person you are talking with the same way. Of course, as you already learned, radio waves are not directional. They disperse in all directions, and anyone with the proper radio receiver can listen in.

You can readily purchase scanners that listen in on analog wireless telephones. In fact, an associate of mine demonstrated just such a thing at a conference once. It was really disturbing to hear folks blathering on their cell phones, oblivious to the fact someone else was listening. Such eavesdropping can be accomplished for less than $100 today. Digital communications has made it more difficult, but it is still possible — they are still radio waves. It just takes more sophisticated gear to accomplish the task.

Eavesdropping on your wireless network is trivial, requiring only a strong antenna, along with the normal wireless networking tools you might have, such as NetStumbler and a packet sniffer. The better the antenna, the easier it is to eavesdrop on someone's network. How much information you get is then a combination of your skill and the degree to which the network is protected using encryption or turnkey vendor solutions.

You always need to be aware of what you are transmitting on your cell or wireless network. If you really don't want it known, then you shouldn't use these technologies without strong encryption. If you think about it, the accidental association we mention above is a form of inadvertent eavesdropping, isn't it?

Man-in-the-middle attacks

So I am standing in the middle of a group, trying to be the man-in-the-middle. It's actually kind of hard, as a group of people is sort of fluid and moves. Man-in-the-middle attacks are the same way: kind of difficult, requiring constant adjustments and an elevated level of knowledge and ability. What is this phenomenon I am discussing?

A *man-in-the-middle attack* is where a rogue agent acts as an access point to the user and as a user to the access point, ending up in the middle of the two ends. All information is then routed through the rogue agent. Man-in-the-middle attacks work in wireless networks in part because 802.1x uses only one-way authentication. There is an implicit trust that the access point you are connecting to is the correct access point. When a man-in-the-middle attack occurs, that trust is abused to trick you into connecting. Your connection is then forwarded to the real access point you wanted to get to, completing your connection and allowing you to go about your business. Meanwhile, all your traffic is being captured and viewed.

Consider doing regular wireless site surveys to see if someone is violating your network by placing unauthorized access points on the network.

Hijacking

Hijacking is similar to the man-in-the-middle-attack. Unfortunately, hijacking is fairly easy to do, especially if users are connecting to a free wireless access point in a hotel or coffee shop.

While sitting in a coffee shop sipping a latte, connect a laptop to the wireless network. Instead of doing the normal activity of opening a browser on the Web, open up a scanning tool to see who else is connected. You might use a security tool called NMAP or one called Look@Lan to see what else is on the network.

After you find some computer addresses, probing them for open ports is easy, and, unless they are running firewall software or intrusion detection, they'll never know. After you locate open ports, it becomes a matter of time to see whether you can access the data on the machine, using open shares they may have left available or a myriad if hacking tools. Most workstations and laptops are poorly secured and therefore fairly vulnerable to attack. Using a free wireless network is one way to be hijacked. There are numerous tools for performing this sort of attack, including:

- ✔ Superscan
- ✔ SNScan
- ✔ Look@Lan
- ✔ Nessus
- ✔ Netcat

Luckily, in the next three chapters, you discover how to secure your network. You need to realize, however, that we do not show you how to protect your *access points* against attacks such as the one we just described. Just the network. You need to look to additional books like *Firewalls For Dummies* by Brian Komar, Ronald Beekelaar, and Joern Wettern (Wiley) or contact us for consulting help. You can reach Barry at lewisb@cerberus-isc.com or Peter at ptdavis@pdaconsulting.com.

Chapter 10

Designing a Secure Network

- -

- -

As we write this book, headlines shout "Wi-Fi's Hot, But Security's Not," "Many Wireless Networks Lack Security," "Under the Radar: Mobile Devices as a Threat to Enterprise," "Wardriver Pleads Guilty in Lowes WiFi Hacks," and "Consultant Finds Security Flaw In Linksys Wireless Router."

On top of these screaming headlines, our experience confirms that less than half of scanned sites have implemented security solutions for their wireless networks. That's a troubling statement, especially given that wireless networking and mobile computing are two of the fastest growing technologies since the Internet. Although wireless networks have ushered in sweeping productivity gains at some enterprises, they have also increased their exposure to security risks in ways that these enterprises do not necessarily understand. In contrast, many organizations are shying away from wireless networking because they perceive wireless security to be as effective as a screen door in a submarine.

But organizations are rapidly deploying wireless devices to increase productivity and improve connectivity. Although wireless computing devices and infrastructure support systems can provide an increase in connectivity, they change the threat mix and also provide increased security risks, resulting in potential vulnerabilities. While we continue to assimilate these technologies into the workplace, we need to ensure that we take a balanced approach regarding the associated vulnerabilities and security risks. You must implement an integrated protection scheme when you deploy wireless technology to support your business and mission-critical operations.

Security as a Cost of Doing Business

There is no denying it: Security represents a cost of doing business. Some of your business is contingent upon secure applications and data. For example, e-business revenue streams may depend on proper security. Security is akin to insurance costs; that is, you pay now to save later. Insurance, after all, is applied risk management. It is reminiscent of the old Fram car filter commercial where the mechanic comments when asked about the price of the filter: "You can pay me now, or you can pay me later." Obviously, the cost of an oil filter is a lot less than the cost of a new engine, but the implementation of some controls now can save you money later.

There is the loss of assets to worry about, but that is not the only concern. Legal actions may result if you fail to meet a general duty of care exhibited as minimum-security standards. Your organization might also have to worry about compliance with specific legislation. In the United States, this could mean

- ✓ **Gramm-Leach-Bliley Act (GLBA):** Protects the privacy of customer information at financial institutions

- ✓ **Health Information Portability and Accountability Act (HIPAA):** Defines standards and procedures for gathering, retaining, and sharing customer information in the healthcare sector

- ✓ **Sarbanes-Oxley Act (SOX):** Affects publicly traded companies governed by the SEC

You might know about other legislation affecting your industry or business. Other countries have or are developing similar legislation. You'll need to know the legal obligations of your particular jurisdiction.

Current resistance to security expenditure will shrink as the information age matures; after all, nobody questions the cost of building security anymore. When we first started in computing, people could not understand the need for passwords, but today, passwords are an accepted control for any system.

In this chapter, we show you how to design a secure network to mitigate the vulnerabilities and security risks introduced by wireless technologies and the infrastructure installed to support them.

Developing a Security Architecture

Although obvious differences exist between wired and wireless networks, the security principles remain the same. By analyzing the security needs of your organization, you can protect it by implementing the right security controls correctly, at the right time. Working in this manner, you can ensure a successful outcome. Developing a security architecture is more important to the security of your organization than any software or hardware you may purchase. Security is not a point product, such as a firewall; rather, it is a process.

Building a secure wireless network is akin to building a house. When you start to build your house, you have to decide whether you will build a ranch, a split-level, a bungalow, a Tudor, a mansard, a neo-classical revival, or what have you. This is your security stance and strategy. Well, the first step in creating a secure wireless network is to determine your stance and establish an enterprise-wide strategy for deployment and usage. Are you a security-conscious organization? Is your industry security-conscious? Do your customers and clients expect secure applications? Do you process or maintain personal information? These are all questions to help you derive your security stance.

At the highest level, your strategy should address the requirements of the following:

- **Confidentiality:** The means for keeping transmitted data secret until it reaches its destination
- **Integrity:** The means by which the recipient of the data transfer can know that the data is intact and that no one has tampered with it
- **Authentication:** Ensures that network access is granted to only approved persons or devices
- **Availability:** The quality of being at hand when needed
- **Accountability:** The responsibility to someone or for some activity

These are high-level goals of your security program. Your strategy should address the following areas as well:

✔ **Determine business needs.** What are the business drivers and needs of your organization? Identify objectives clearly, and make sure that the benefits outweigh the risks.

✔ **Integrate wireless policies into existing IT policies.** Remember that wireless solutions are an extension of the wired network.

✔ **Clearly define wireless network ownership.** This ensures control as well as response when you identify security threats. Also, defining network ownership should nip backdoor or rogue access points in the bud.

✔ **Protect the existing infrastructure.** This is what it really is about. Do not place wireless devices directly on the internal network. Instead, provide a separate network or demilitarized zone to control access to the wired network.

✔ **Educate users about wireless policies.** This includes providing awareness sessions for employees.

Your policy should consider the assets you intend to protect: sensitive data and network services. You cannot develop policy statements without considering the threats you are trying to prevent: equipment damage or theft, denial of service, unauthorized access, fraud, data theft, personal information exposure, data insertion, and legal liabilities.

To build a house, you need a blueprint or architectural plan. The blueprint lays out what is expected. You want to know how many washrooms or other services you will have. Likewise, your security architecture should have a plan or blueprint. A good security plan includes the following network services:

✔ **Authentication:** One entity (that is, simply a person or system) proves to the other its identity.

✔ **Access control:** You allow or deny an entity access to the network.

✔ **Replay prevention:** An entity can determine a previously sent message.

✔ **Message integrity:** An entity can verify that no one has changed the content of a message in transit.

✔ **Message privacy:** Sensitive information is encrypted when transmitted between two wireless entities to prevent interception and disclosure or to prevent a third party from tracking communications between two other entities.

✔ **Non-repudiation:** An entity can verify the origin or the receipt of a specific message.

✔ **Accountability:** An entity can trace the actions of an entity uniquely to that entity.

✔ **Key protection:** The system can protect the confidentiality of a key used by an entity.

When building a home, you want to ensure that you begin with a strong foundation. You pour some concrete and form the basement or foundation. The foundation or baseline of any security architecture is the security policy.

Developing a Wireless Security Policy

Again, using the house analogy, before you put the shovel in the ground, a surveyor comes to the property and maps its dimensions. Well, before you start your site survey policy, you need to map your organization's security policy. Develop a security policy that addresses the use of wireless technology, including 802.11, 802.15, 802.16, and 802.20. A security policy is the foundation on which you rationalize and implement other countermeasures — the operational and technical ones. A documented security policy allows your organization to define acceptable architecture, implementation, and uses for wireless technologies. It answers the big questions. Do you allow ad hoc mode? Do you allow departments or units to install access points? Do you allow WPAN, WLAN, or WWAN technology? There are many formal statements that you need to make in your policy.

Keep in mind that policies are mandated specifications or operations. They deal with the assessment of risk within your organization. You cannot write effective policy until you understand the risks to your organization and how you intend organizationally to deal with those risks. These policies provide the basis for operations — and consequently, the basis of compliance reviews.

Policies are not stagnant; you don't write them in stone. Your internal and external environments are constantly changing — as a result, so are the threats and the attendant risks. You should periodically review and update your policy to address technology improvements that may provide practical application for your organization without introducing additional security risks and vulnerabilities. Determine what works and why it works. Determine what doesn't work and why it doesn't. Make sure that your policy is current by rewriting any dysfunctional or archaic policies. This is a constant and cyclic process as your organization moves forward.

It is important to remember the intended audience when you draft security policy. Don't make the policy too difficult to read or comprehend. It is important to create formal policy to minimize potential confusion and to clear up any ambiguity. Sometimes your policy only formalizes the status quo. Make sure that your policies are relevant. If people don't understand what the policy means to them, they will disregard it. Make your policy succinct but precise.

Every organization has or should have a format or template for policy, so we won't tell you how to write your policy. However, you should consider the following topics:

- **Purpose:** Tell the audience that the policy applies to wireless networks.

- **Scope:** The policy may apply only to WLANs, but it may also apply to all wireless technology, so you need to specify the scope. People can read the scope and decide whether their network is in or out of scope and whether the policy applies to them.

- **Policy:** State the policy very clearly. A good policy document is usually a maximum of three pages long. Generally, you specify the clients' rights and responsibilities and expected actions or behaviors. Many organizations include standards and procedures in their policy, which you shouldn't do. If you're not sure of the difference, you can refer to ISO 17799 (www.iso17799.net), which tells you about the many tiers of documentation.

- **Enforcement:** You should state what someone can expect in the way of sanctions should they not comply with your policy. Your legal counsel should review this section carefully (well, actually, the whole document).

- **Exceptions:** Circumstances may arise in which, for one reason or another, someone cannot comply with the policy. You should have a mechanism in place to handle the reporting and approval of any exceptions.

- **Definitions:** You cannot (and should not) expect every reader of your document to understand all the technical jargon. You should write the policy so the average layperson can understand it. Where jargon is unavoidable, you may need to define some terms.

- **Document history:** Whether you put it at the beginning or the end is inconsequential, you must include a document history. The reader should have the trail of revisions for the document.

After the powers that be approve the policy, make sure that everyone gets a copy. If need be, make sure that everyone understands the policy. We often see gaps in security programs in which people who have the necessary skills want to do the right thing but don't understand how. Tell them. After you tell them, get them to sign a document saying they read and understood the policy. It's also a worthwhile idea to reaffirm annually that they understand the policy.

Every organization is different, but some typical security policy topics are

✔ Use of default SSIDs, encryption keys, and passwords

✔ Trust level of base station, bridges, and clients

✔ Access control method(s): MAC ACLs, 802.1x, and SSL

✔ Method for configuration changes: console ports, TFTP, Telnet, HTTP, and HTTPS

✔ Access policies for authorized APs, stations, groups, users, and guests

✔ Authentication credentials

✔ Authentication method(s): none, shared key, EAP, VPN, and SSL login

✔ Encryption technology: 802.11, WEP, WPA, AES, network, transport, and application

✔ Required software and settings for AP, authentication servers, and clients (including firewall and antivirus)

✔ Filtering: MAC, protocol, and watch lists

If you're having trouble starting your wireless security policy, the U.S. Department of Defense has a document at www.defenselink.mil/nii/ org/cio/doc/it-wireless-policy-092502.pdf that may give you some thoughts. The folks at the SysAdmin, Audit, Network, Security (SANS) Institute offer some excellent resources for policy, standard and guidelines implementation at www.sans.org/resources/policies. You'll find they even have a wireless communication policy template that you can download in Word format. The U.S. Defense Information Systems Agency (DISA) also has policy references at http://iase.disa.mil/policy.html#wireless.

Developing Wireless Security Standards

As stated earlier, many organizations include standards and procedures in their policy. You should not. Even so, you need to set some measures. Some wireless security standards that you may develop include

✔ **Standard support:** Do you support 802.11a, b, or g?

✔ **Equipment:** Do you support equipment from any vendor?

✔ **Hours of operation:** Do you allow off-hours connections?

- ✔ **Naming standard:** How do you name access points and bridges?

- ✔ **Channel support:** What channels do you use?

- ✔ **Data rates:** What data rates do you support?

- ✔ **Performance:** What are traffic thresholds, and how many stations do you allow an access point to support?

- ✔ **Encryption algorithm:** What algorithm does your organization support?

- ✔ **Key lengths:** What is the minimum key length?

- ✔ **Extensible Authentication Protocol:** What flavor of the many types of EAP do you support?

- ✔ **Password:** What is the password length, and how often do you change it?

- ✔ **Upgrades:** When do you apply upgrades or patches?

- ✔ **Tunneling protocol and algorithms:** What layer and what algorithm?

- ✔ **Key distribution and refresh procedures:** How do you disseminate keys?

These are just some of the topics to cover in your standards. Think of the definition of *standard.* It is the required degree or level of requirement, excellence, or attainment. The standard is the ideal. Your management and internal and external auditors will measure you on how well you meet the established standards.

Developing Wireless Security Best Practices

You do need to develop policies, standards, and practices for your organization, but you may find it useful to base these on best practices. We state earlier that best practices demonstrate prudence. There is no agreement yet on the required set of standards for secure wireless access points, but you can find agreement on best practices. To protect a WLAN from attack, enterprises need to be up-to-date with their security best practices. These should include the best practices covered in the following sections.

General best practices

- ✔ Designate an individual to track the progress of 802.11, 802.15, and 802.16 security products and standards (IETF, IEEE, etc.) and the threats and vulnerabilities with the technology. (See Chapter 16.)

✔ Keep your computers and Wi-Fi devices powered up at all times, but power-down your broadband modem afterhours. (See Chapter 10.)

✔ Ensure that wireless networks are not used until they comply with the security policy. (See Chapter 10.)

✔ Complete a site survey to measure and establish the AP coverage for the agency. (See Chapter 2.)

✔ Ensure that the ad hoc mode for 802.11 has been disabled unless the environment is such that the risk is tolerable. (See Chapter 6.)

✔ Enable all security features of the WLAN product. (See Chapter 11.)

Access point best practices

✔ Maintain a complete inventory of all APs and wireless devices. (See Chapter 16.)

✔ Control the broadcast area through cell sizing. Many wireless access points let you adjust the signal strength. (See Chapter 18.)

✔ Place your access points as far away as possible from exterior walls and windows. Place them in the interior of the building where appropriate. (See Chapter 2.)

✔ Place APs in secured areas to prevent unauthorized physical access and user manipulation. (See Chapter 2.)

✔ Mount your access points out of reach and out of plain view. Bolt them down or secure them in locked steel enclosures. (See Chapter 5.)

✔ Test the signal strength. (See Chapter 16.)

✔ Make sure that you use the reset function on APs only when needed and that it can be invoked only by someone in an authorized group of people. (See Chapter 5.)

✔ Restore the APs to the latest security settings when someone uses the reset function. (See Chapter 5.)

✔ For 802.11b and g devices, ensure that AP channels are at least five channels apart from any other nearby wireless networks to prevent interference. Use 802.11a when you need more co-located APs. (See Appendix C.)

✔ Understand and make sure that all default parameters are changed. (See Chapter 5.)

✔ Disable all nonsecure and nonessential management protocols on the APs. If you have Cisco devices, disable Cisco Discovery Protocol (CDP) when not needed. (See Chapter 18.)

✔ When disposing of access points that will no longer be used by the organization, clear access point configuration to prevent disclosure of network configuration, keys, passwords, and so on. (See Chapter 10.)

✔ If the access point supports logging, turn it on and review the logs on a regular basis. (See Chapter 16.)

Password best practices

✔ Be sure to change the default password on all access points. (See Chapter 5.)

✔ Use a strong password to protect each access point. (See Chapter 5.)

✔ Ensure that all passwords are changed regularly. (See Chapter 5.)

SSID best practices

✔ Use SSID (Service Set Identifier) wisely. Don't use the default and don't use the name of your company as the SSID. (See Chapter 18.)

✔ Buy access points that let you disable SSID broadcasting. This prevents access points from broadcasting the network name and associating with clients that aren't configured with your SSID. (See Chapter 18.)

✔ Immediately change an access point's default SSID. (And while you're at it, change the default username and administrator password, too.) (See Chapter 18.)

Authentication best practices

✔ Implement user authentication. Require access point users to authenticate. (See Chapter 11.)

✔ Upgrade access points to use implementations of the WPA and 802.11i standards. Also, as you implement user authentication on the access points, reuse any existing servers that provide authentication for your other network services, such as RADIUS. (See Chapter 11.)

✔ Use MAC (Media Access Control) address authentication where practical. When you have a manageable number of wireless users and just a few access points, MAC addressing lets you restrict connections to your access points by specifying the unique hardware address of each authorized device in an access control list and allowing only those specific devices to connect to the wireless network. (See Chapters 11 and 18.)

✔ Enable user authentication mechanisms for the management interfaces of the AP. (See Chapter 5.)

Encryption best practices

✔ Secure the WLAN with IPSec VPN technology or clientless VPN technology. (See Chapter 12.)

✔ Turn on the highest level of security your hardware supports. Even if you have older equipment that supports only WEP, ensure that you enable it. Whenever possible, use at least 128-bit WEP. (See Chapters 11 and 18.)

✔ Ensure that encryption key sizes are as long as possible. (See Chapter 11.)

✔ Make sure that default shared keys are periodically replaced by more secure unique keys. (See Chapter 11.)

Client best practices

✔ Deploy personal firewalls and virus protection on all mobile devices. (See Chapter 8.)

✔ Ensure that the client wireless adapter and AP support firmware upgrades so that security patches may be deployed as they become available. (See Chapter 6.)

✔ Ensure that users on the network are fully trained in security awareness and the risks associated with wireless technology. (See Chapter 10.)

✔ Regularly scan for rogue access points on the network by using a wireless scanner or a packet analyzer. (See Chapter 16.)

✔ Use antivirus software on all wireless clients. (See Chapter 10.)

✔ Use personal firewall software on all wireless clients. (See Chapter 10.)

✔ Use a secure transport for wireless communications: for example, IPSec, SSL, or SH. (See Chapter 12.)

✔ Disable WNIC when not used. (See Chapter 10.)

✔ Update and enable client security software and patch OS. (See Chapter 6.)

✔ Take regular backups. (See Chapter 10.)

Network best practices

✔ Deploy enterprise-class protection technologies. This includes employing a firewall on the demilitarized zone and client firewalls on every desktop; VPN services that encrypt all traffic to and from wireless devices; wireless and network intrusion detection systems; antivirus software for the network, server, and desktop; regular vulnerability assessments of the WLAN; and policy compliance tools. (See Chapter 18.)

✔ Install a properly configured firewall between the wired infrastructure and the wireless network. (See Chapter 18.)

✔ Use bridges, switches and gateways to segment the network. (See Chapter 14.)

✔ Use Layer 2 switches in lieu of hubs for AP connectivity. (See Chapter 14.)

✔ Do not connect wireless access points to hubs. (See Chapter 18.)

✔ Disable DHCP. (See Chapter 18.)

✔ Ensure that management traffic destined for APs is on a dedicated wired subnet. (See Chapter 13.)

✔ Configure SNMP settings on APs for least privilege (that is, read only). (See Chapter 18.)

✔ Disable SNMP if it is not used. SNMPv1 and SNMPv2 are not recommended. Use SNMPv3 and/or SSL/TLS for Web-based management of APs. (See Chapter 18.)

✔ Use a local serial port interface for AP configuration to minimize the exposure of sensitive management information. (See Chapters 5 and 18.)

✔ Deploy intrusion detection agents on the wireless part of the network to detect suspicious behavior or unauthorized access and activity. (See Chapter 16.)

✔ Use static IP addressing on the network. (See Chapter 18.)

✔ Perform comprehensive security assessments at regular and random intervals (including validating that rogue APs do not exist in the 802.11 WLAN) to fully understand the wireless network security posture. (See Chapter 16.)

✔ Turn off communication ports during periods of inactivity when possible. (See Chapter 4.)

Ensure that all users on the network are fully trained in computer security awareness and the risks associated with wireless technology. A security awareness program helps users establish good security practices to prevent inadvertent or malicious intrusions into an organization's information systems.

Managing Your Wireless Security Policy

Chapter 16 covers wireless packet analyzers and intrusion detection systems. These products can help you manage your policy. You can set a trap for unencrypted traffic should you have a policy that requires encrypted wireless traffic. Or you can look for rogue or unauthorized access points. Other products that can help you manage your wireless policy include

- **AirDefense** (www.airdefense.net/products/features/policy.html): Monitor and enforce configuration policies, WLAN device and roaming policies, performance policies, channel policies, and vendor policies.

- **AirWave Management Platform** (www.airwave.com): Automatic configuration, policy push, and compliance auditing.

- **Chantry BeaconWorks** (www.chantrynetworks.com): Allows central management of APs.

- **Cirond Winc Manager** (www.cirond.com): WEP key distribution, location-based access control, provisioning system, and real-time mapping.

- **Computer Associates Unicenter Wireless Site Management** (www.ca.com): Key management, wireless rogue AP detection, provisioning, network discovery and mapping, and WAP configuration and administration.

- **Enterasys Secure Networks** (www.enterasys.com/solutions): Provisioning and security solutions.

- **Sygate** (www.sygate.com/products/enterprise_policy_management.htm): Definition of policies based on client behavior.

- **Vernier System 6500** (www.verniernetworks.com/products/control.htm): Hardware enterprise-class WLAN gateway solution that provides network administrators with the flexibility of configuring access rights based on user, group, time, and location (centralized policies).

- **Wavelink Mobile Manager** (www.wavelink.com): WEP key management, enterprise ACLs, and policy push.

Designing a Secure Network

When you design a secure network, remember to design a system in depth. Figure 10-1 can serve as a model for this concept. At the center of the "security onion" is the data that you intend to protect. Your security design should consider personnel, administrative, operational, software, and hardware security measures. It starts with hiring trustworthy individuals. Using multiple layers of security, such as WPA, 802.1X, and IPSec provides high levels of security but introduces complexity and increases costs. So do your homework, and implement only the necessary and sufficient control mix.

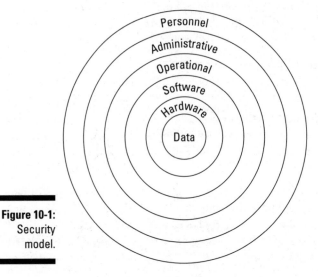

Personnel
Administrative
Operational
Software
Hardware
Data

Figure 10-1:
Security
model.

Performing a Risk Analysis

Unless you are the owner, CEO, or head of your organization, one day someone will probably say to you, "I need some solid evidence that your security programs are contributing to the organization's productivity, its competitiveness, and ultimately its bottom line." When you are asked these questions, you better know the following:

- ✔ How vulnerable is the organization to known attacks?
- ✔ When was the analysis last done?
- ✔ What percentage of company software, people, and supplies has been reviewed for security issues?

✔ What percentage of critical data is strongly protected?

✔ What percentage of downtime results from security problems?

✔ What percentage of nodes in the network does IT manage?

You should perform a risk assessment to understand the value of the assets that need protection in your organization. Your management wants to know the threats and risks associated with today's networks and the method for controlling them. Security and controls improve quality and performance, which are the keys to success in any organization. So, you should agree that you should have security and controls. Saying this another way, you should manage risks. But what does *manage risks* mean? Risk management is the optimized allocation of limited resources to

✔ Mitigate risks

✔ Transfer risks

✔ Recover from risk events

Your organization will perform better when you manage risks, which means more effective use of resources, more responsiveness to clients, and compliance with laws. So what's the problem? Why not just find and fix all your risks? Because perfect security is infinitely expensive! No organization — not even the government (or especially the government) — has unlimited resources.

You must measure risk. You can use *High, Medium,* and *Low,* but this is a difficult sell when you go to the boss and say, "Hey, boss, I need high dollars to manage this high risk!" What is the likely response? "I need better data than that!" Therefore, you must measure your risks and not merely express your opinions. You can calculate *expected loss,* which is the stream of risk losses expressed quantitatively, that you could reasonably expect to experience in the future. Some organizations do this by measuring the return on investment (ROI). Typically, ROI is a measure of an organization's performance. It is finite: total capital divided into income. Normally, ROI is defined by the business as an incremental gain on an action. There are three ways to maximize ROI:

✔ Minimize costs

✔ Maximize returns

✔ Accelerate the timing of returns

Alternatively, you could calculate the *Return on Security Investment* (ROSI), which is normally defined as the value of loss deference or reduction to dollars invested on security controls. It is indefinite: It has no exact limits. Some security investments have specific ROI, such as provisioning users or corporate insurance, but most don't. ROSI is an incremental gain on an action. There are four ways to maximize ROSI:

- ✔ Minimize/eliminate operational losses
- ✔ Minimize investment
- ✔ Maximize positive returns (where ROI applies)
- ✔ Accelerate the timing of returns

Your goal is to implement cost-effective security, in which the expected cost of a control is less than the expected loss. Such controls generate a positive ROSI; that is, you can expect to save money over time. Ideally, you want to deploy the most cost-effective controls — those that maximize ROSI. Your challenge is to measure ROSI for given security controls. You should try to base measurements on empirical data and mathematical analysis, rather than opinions. You should evaluate all proposals, techniques, products, and services in terms of ROSI. You should establish best practices based on ROSI. Unfortunately, most companies currently base security decisions on expert opinion and conventional wisdom, not on empirical data and mathematical analysis.

Perform a risk assessment to understand the value of the assets in your organization that need protection. Understanding the value of organizational assets and the level of protection required is likely to enable more cost-effective wireless solutions that provide an appropriate level of security. You don't want to spend money to protect data that has no value. We doubt that you will find any case in which the data has no value, but you don't want to spend more on security measures than the value of the data.

Several companies sell risk management software, including Methodware Enterprise Risk Assessor (www.methodware.com) and Risk Services & Technology RiskTrak (www.risktrak.com).

Chapter 11

Maintaining Network Security

· ·

· ·

*I*n this chapter, we look at several built-in security features of 802.11 for network security. Risks in wireless networks are equal to the sum of the risk of operating a wired network (as in operating a network in general) plus the new risks introduced by weaknesses in wireless protocols.

In Chapter 2, we discuss the need to specify security requirements. This includes determining the security stance of the organization. You need to perform a security assessment prior to implementation to determine the specific threats and vulnerabilities that wireless networks will introduce in your environment. In performing your assessment, you should consider your existing security policies, known threats and vulnerabilities, legislation and regulations, safety, reliability, system performance, the life-cycle costs of security measures, and technical requirements. After you complete your risk assessment, you can begin planning and implementing the measures that you will put in place to safeguard your systems and lower your security risks to an acceptable level. Your organization should periodically reassess the policies and measures that it puts in place because technologies and malicious threats are ever-changing. As with wired networks, you must make your management aware of security issues.

Understanding Security Mechanisms

The IEEE 802.11 specification identified several features to provide a secure operating environment. Your challenge is to decide how many of these security features you need. In this chapter, we provide an overview of the inherent network security features to better illustrate the limitations. When reviewing the security requirements, we use the following requirements:

- ✔ **Authentication:** One entity proves to the other their identity.

- ✔ **Access control:** An entity can be allowed or denied access to the network.

- ✔ **Replay prevention:** An entity can determine a previously sent message.

- ✔ **Message integrity:** An entity can verify that no one has changed the content of a message in transit.

- ✔ **Message privacy:** Sensitive information is encrypted when transmitted between two wireless entities to prevent interception and disclosure or prevent a third party from tracking communications between two other entities.

- ✔ **Non-repudiation:** An entity can verify the origin or the receipt of a specific message.

- ✔ **Accountability:** An entity can trace the actions of an entity uniquely to that entity.

- ✔ **Key protection:** The system can protect the confidentiality of a key used by an entity.

As we go through this chapter, you will note that the 802.11 standard did not specifically address these security services. The 802.11 standard attempts to address privacy and integrity but falls well short and does not offer the other security services.

As with many newer technologies (and some older ones), you may not find the available security features as comprehensive or robust as you would like. Although the security features have weaknesses described as you will see in this chapter, they can provide a degree of protection against unauthorized disclosure, unauthorized network access, and other active probing attacks. We strongly recommend that you use the built-in security features as part of an overall defense-in-depth strategy. Unfortunately, vendors frequently disable the built-in security features by default. You must enable, use, and routinely test the built-in security features, such as authentication and encryption, that exist in wireless technologies.

Three States of Authentication

A necessary security service is authentication. It is as basic a service as you can get. In the standard 802.11, we don't authenticate users. If you want, you can make sure someone knows the shared key. Before we finish this chapter, we will show you why you don't want to use the shared key to authenticate.

While authenticating, a wireless client goes through three states:

- **Unauthenticated and unassociated:** The client selects a basic service set by sending a probe request to an access point with a matching SSID.

- **Authenticated and unassociated:** The client and the access point perform authentication by exchanging several management frames. After authentication, the client moves into this state.

- **Authenticated and associated:** Client must send an association request frame, and the access point must respond with an association response frame.

A client can authenticate to many access points, but will associate only with the access point with the strongest signal.

In the second state, we just casually mention the client authenticates to the access point. It's not quite that simple.

Authentication

The IEEE 802.11 specification defines two ways to "validate" wireless users attempting to gain access to a wired network: open system authentication and shared-key authentication. Shared-key authentication is based on cryptography, and the other is not. The open system authentication technique is not truly authentication; the access point accepts the mobile station without verifying the identity of the station.

With open system authentication, the AP authenticates a client when the client simply responds with a MAC address during the two-message exchange. The open system authentication process is as follows:

1. **Client makes a request to associate to an access point.**

2. **AP authenticates client and sends a positive response and client is associated.**

Shared-key authentication is a cryptographic technique for authentication. It is a simple "challenge-response" scheme based on whether a client has knowledge of a shared secret. In this scheme, the access point generates a random 128-bit challenge and sends it to the wireless client. The client, using a cryptographic key that is shared with the access point, encrypts the challenge, or *nonce* (as it is called in security vernacular), and returns the result to the AP. The AP decrypts the result computed by the client and allows access only when the decrypted value is the same as the random challenge transmitted. The algorithm used in the cryptographic computation and for the generation of the 128-bit challenge text is the same RC4 stream cipher used for Wireless Equivalent Privacy (WEP).

This authentication method is a rudimentary cryptographic technique that does not provide mutual authentication. That is, the client does not authenticate the AP, and therefore there is no assurance that a client is communicating with a legitimate AP and wireless network. It is also worth noting that simple unilateral challenge-response schemes have long been known to be weak. They suffer from numerous attacks, including the infamous "man-in-the-middle" attack. The shared-key authentication process follows:

1. **Client requests association.**

2. **AP sends random cleartext (128-bit challenge).**

3. **Client encrypts challenge.**

4. **AP verifies the challenge.**

5. **The access point authenticates the client and sends a positive response and then associates the client.**

Table 11-1 lists the pros and cons of the two types of authentication. The IEEE 802.11 specification does not require shared-key authentication.

Table 11-1   Open System versus Shared-Key Authentication

Open System	*Shared-Key*
A station is allowed to join a network without any identity verification.	A station is allowed to join the network when it proves it shares the WEP key.
1-stage challenge/response (not required).	2-stage challenge/response (required).
Non-cryptographic.	Cryptographic using RC4.

Logically, you may guess that shared-key authentication is more secure than open system authentication. But this is not the case. Because of the way the shared-key authentication is done, it is less secure. Let's look at why. An attacker gathers management messages from the authentication process. One message contains the random challenge in cleartext. The next message contains the encrypted challenge using the shared-key. The encryption process is simple. The algorithm does an exclusive OR on the plaintext to derive ciphertext as follows:

```
P XOR R = C
```

From here, the rest is just simple math:

```
If P XOR R = C then C XOR R = P
If P XOR R = C then C XOR P = R
```

Now, the attacker knows everything from passive networking monitoring: algorithm number, sequence number, status code, element ID, length, and challenge text. The attacker requests authentication. The access point responds with a cleartext challenge. The attacker uses the challenge with the value R above to compute a valid authentication response frame by XORing the two values together and computes a valid CRC value. Finally, the attacker responds with a valid authentication response message and associates with the AP to join the network. Because of the flaw, the attacker did not need to know the shared-key!

Protecting Privacy

The 802.11 standard supports privacy (confidentiality) through the use of cryptographic techniques for the wireless interface. The WEP cryptographic technique for confidentiality also uses the RC4 symmetric-key, stream cipher algorithm to generate a pseudo-random data sequence. This *key stream* is simply added modulo 2 (exclusive ORed) to the data to be transmitted. Through the WEP technique, you can protect data from disclosure during transmission over the wireless link. WEP is applied to all data above the 802.11 WLAN layers to protect datagrams such as Internet Protocol (IP) and Internet Packet Exchange (IPX), or application protocols such as HyperText Transfer Protocol (HTTP) and Simple Mail Transfer Protocol (SMTP).

As defined in the 802.11 standard, WEP supports only a 40-bit cryptographic key size for the shared key. However, numerous vendors offer nonstandard extensions of WEP that support key lengths from 40 bits to 104 bits. At least one vendor supports a key size of 128 bits (that is, 152 bits). The 104-bit WEP key, for instance, with a 24-bit initialization vector (IV) becomes a 128-bit RC4 key. In general, all other things being equal, increasing the key size increases the security of a cryptographic technique. However, it is always possible for flawed implementations or flawed designs to prevent long keys from increasing security. Research has shown that key sizes of greater than 80 bits, for robust designs and implementations, make brute-force cryptanalysis (code breaking) an impossible task. For 80-bit keys, the number of possible keys — a key space of more than 10^{26} — exceeds contemporary computing power. In practice, most WLAN deployments rely on 40-bit keys. Moreover, recent attacks have shown that the WEP approach for privacy is, unfortunately, vulnerable to certain attacks regardless of key size. The attacks mentioned above are described later in the following sections.

Protecting Message Integrity

The IEEE 802.11 specification also outlines a way for providing data integrity for messages transmitted between wireless clients and access points. This security service was designed to reject any messages that an active adversary "in the middle" had changed. This technique uses a simple Cyclic Redundancy Check (CRC) approach. The access point and client compute a CRC-32 or frame check sequence called an integrity check value (ICV) for each frame prior to transmission. Referring to Figure 11-1 (later in the chapter), you can see that WEP then encrypts the integrity-sealed packet using the RC4 key stream to provide the ciphertext message. The receiver decrypts the frame and recomputes the CRC on the message. The CRC computed at the receiving end is compared with the one computed with the original message. When the CRCs are not equal, there is an error, and the receiver discards the frame. Great idea, but again poorly implemented. It is possible to flip bits and still end up passing the CRC check. The CRC is not a cryptographically secure mechanism such as a secure hash, message digest, or message authentication code (MAC).

CRC-32 and other linear block codes are inadequate for providing cryptographic integrity. Message modification is possible. Linear codes are inadequate for protecting against intentional data integrity attacks. You need real cryptographic protection to prevent deliberate attacks. Use of non-cryptographic protocols often facilitates attacks against the cryptography. In our case, it does. One reason is that we use our 64- or 128-bit key for integrity and privacy, a cryptography no-no.

Filtering the Chaff

As mentioned previously, we want to build our security in-depth. We never rely on one control because it may fail. You can build defense-in-depth by using some of the filtering capabilities offered on your access point. They are not the strongest and you should not rely on only these filters, but they may act as a departure point for your network security.

SSID filtering

The simplest filter you have is SSID filtering. You can eliminate casual attempts to join your network by turning off SSID broadcast and requiring your client to know the SSID of the network. Let's be sure we understand that an SSID is not a passcode of any kind but an identifier for your network. Now, you can use Kismet, Wellenreiter, and other tools to monitor packets until you figure out the SSID, so this might discourage an individual looking for the "low hanging fruit," but not a determined attacker.

MAC filtering

MAC (or physical or hardware) address filtering provides basic control over the stations that you want connecting to your access point. A MAC (media access control) address is a hardware or physical address uniquely identifying each computer or attached device on a network. It is a 48-bit number set by the manufacturer. The 48 bits break down into a 24-bit organizationally unique identifier (OUI), assigned by the IEEE, and a 24-bit unique card identifier. You can find a list of OUIs at `http://standards.ieee.org/regauth/oui/index.shtml`. The address is a unique 6-part hexadecimal with each part numbered from 00 to FF. You can write the address unhyphenated (for example, 123456789ABC) or with one hyphen (for example,123456-789ABC), but correctly you should write it hyphenated by octets (for example, 12:34:56:78:9A:BC). The numbering scheme gives a theoretical 281,474,976,710,656 addresses — more than 56,000 MAC addresses for each person on the planet! However, the flat addressing scheme limits the available addresses to 2^{24} for each vendor. Because we don't have 2^{24} vendors, some addresses are wasted. When sending a frame, you send the frame to the hardware address ultimately. You use software addresses (for example, IP addresses) to route packets to the destination subnet or segment.

You can use the MAC address to restrict access based on MAC access control lists (ACLs) that are stored and distributed across many APs, although some other access points have only the ability to filter trusted MAC addresses. Regardless, the MAC filter grants or denies access to a computer using a list of permissions designated by MAC address.

The Ethernet MAC filter, however, does not represent a strong defense mechanism by itself. Because your client transmits its MAC address in the clear, someone can easily capture the MAC address. Malicious users can spoof a MAC address by changing the actual MAC address on their computer to a MAC address that has access to the wireless network. You can add a *NetworkAddress* to the Registry with regedit. (Don't forget to back up your registry before making changes to any registry entry.) Alternatively, you can use the Set MAC Address software (www.klcconsulting.net) shown in Chapter 17. If you are using UNIX/Linux, use the ifconfig tool or a short C program calling the ioctl() function with the SIOCSIFHWADDR flag. You can also find a program called macchanger to help out. For the Mac OS X platform, use xnu (www.securemac.com/macosxxnu.php) or etherspoof (http://slagheap.net/etherspoof).

Because someone can use a tool like SMAC to change her MAC address to any value, this may negate the value of MAC filtering. It may have some value against casual eavesdropping, but it is not effective against determined adversaries. However, you should weigh the administrative burden of enabling the MAC ACL (assuming they are using MAC ACLs) against the true security provided. In a medium-to-large network, you may find the burden of establishing and maintaining MAC ACLs or filters exceeds the value of the security countermeasure. In addition, most products support only a limited number of MAC addresses in the MAC ACL or filter.

You may find the size of the access control list insufficient for medium-to-large networks. You also may find this feature difficult to implement in a dynamic environment: Configuring your access points for each and every trusted client can be quite tedious. Table 11-2 shows the pros and cons of MAC Filtering.

Table 11-2	MAC Filtering
Pros	*Cons*
Predefined users accepted	Administrative overhead
Filtered MACs do not get access	Cost of implementation
Provides a good first level of defense	Administrative nightmare

You may find that enabling this security feature is more effort than the actual security benefit that it provides. For small networks where you have fewer than ten workstations, MAC filtering might prove practicable. Some security professionals believe that you don't need both MAC filtering and shared-secret authentication since they basically accomplish the same thing.

Protocol filtering

Although not specified in the 802.11 standard, some vendors have provided protocol filtering. Like MAC filtering, this is another way to minimize risk. You can specify inbound and outbound allowable protocols. You must take care when setting up protocol filtering, or you may find you have blocked clients or let everyone in. You can use protocol filtering to prevent anyone from trying to use the Simple Network Management Protocol (SNMP) to reconfigure your AP. Similarly, you can filter Internet Control Message Protocol (ICMP) messages and potentially prevent some denial-of-service (DoS) attacks. The benefits are great and the disadvantages are small: potentially locking out authorized clients. You're best to use protocol filtering to block unwanted traffic.

Some vendors also offer port forwarding. Port forwarding associates traffic destined for a specific port to a device on the internal network that you cannot necessarily access from the outside. This is another useful security feature that you should use to your advantage.

Using Encryption

The three basic security services defined by the IEEE 802.11 standard are as follows:

✓ **Authentication:** A primary goal of WEP was to provide a security service to verify the identity of communicating client stations. This provides access control to the network by denying access to client stations that cannot authenticate properly. This service addresses the question, "Are only authorized persons allowed to gain access to my network?"

✓ **Integrity:** Another goal of WEP was a security service developed to ensure that messages are not modified in transit between the wireless clients and the access point in an active attack. This service addresses the question, "Is the data coming into or exiting the network trustworthy — has it been tampered with?"

✔ **Confidentiality:** Confidentiality, or privacy, was a second goal of WEP. It was developed to provide the "privacy achieved by a wired network." The intent was to prevent information compromise from casual eavesdropping (passive attack). This service, in general, addresses the question, "Are only authorized persons allowed to view my data?"

The first two items in the preceding list are covered previously in this chapter. We use shared-key or open system for authentication and CRC-32 for frame integrity. It is now time to tackle the issue of confidentiality. The popular press has done a lot to discourage organizations and individuals from using wireless networks. If you have been paying attention, then you are aware of all the negative articles about wireless security, especially encryption. Part of the problem is that people (including the press pundits) don't understand the basis for WEP. As implied by its name, the developers of Wired Equivalent Privacy intended that it give clients the same level of security found on a wired network (which, quite frankly, isn't much). Except for a fully switched environment, eavesdroppers have their way with packets traversing a wired network. WEP was never intended to provide message integrity, non-repudiation, and confidentiality. We will explain some of the shortcomings of the WEP algorithm in this chapter.

Hip to WEP

WEP is a shared key only. It uses the symmetrical RC4 (Ron's Code 4) algorithm and a PRNG (Pseudo-Random Number Generator). The original standard specified 40- (a.k.a. 64) and 128-bit key lengths, with a 24-bit initialization vector (IV). WEP encrypts layers 3 through 7, but does not encrypt the MAC layer (that is, layer 2). Each client has the keys and other configuration data. We know that there is nothing wrong with the RC4 algorithm. After all, it is used in your browser for Secure Sockets Layer (SSL). The problem is in the implementation of the algorithm

Figure 11-1 shows the WEP encryption process. The purpose of WEP is to encrypt a plaintext message. So, that is where the process begins. WEP performs a 32-bit cyclical redundancy check (CRC) checksum. In WEP terms, this is the integrity check value (ICV), which is concatenated to the end of the plaintext message. We take the secret key and concatenate it to the initialization vector (IV). Plug this secret key-IV combination into the RC4 PRNG and output the key stream sequence. The key stream is a bit stream (0s and 1s) equal in length to the plaintext message plus CRC combination. Finally, we perform an exclusive OR (XOR) operation between the plaintext message plus CRC combination and the key stream. The result is the ciphertext. WEP prepends the IV (unencrypted) to the ciphertext and includes it as part of the transmitted data.

You can find out more about CRC at `www2.rad.com/networks/1994/err_con/crc.htm`.

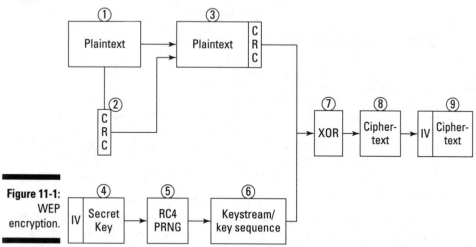

Figure 11-1:
WEP
encryption.

Huh? Perhaps walking through the decryption process will help. The algorithm takes the IV, which is in plaintext, and prepends it to the secret key, which the decrypter knows. WEP then plugs the result into the RC4 to regenerate the key stream. Next, the algorithm XORs the key stream with the ciphertext, which should give us the plaintext value. Finally, WEP re-performs the CRC-32 check-sum on the message and ensures that it matches the integrity check value in our encrypted plaintext. Should the checksums not match, WEP assumes that someone tampered with the packet, and will discard it.

As mentioned previously, access points generally have only three encryption settings available: none, 40-bit shared key, and 104-bit setting. The setting of *none* represents the most serious risk because someone can easily intercept, read, and alter unencrypted data traversing the network. A 40-bit shared key will encrypt the network communications data, but there is still a risk of com-promise. The 40-bit encryption has been broken by brute force cryptanalysis using a high-end graphics computer and even low-end computers; conse-quently, it is of questionable value. In general, 104-bit encryption is more secure than 40-bit encryption because of the significant difference in the size of the cryptographic key space. Although this is not true for 802.11 WEP because of poor cryptographic design using IVs, it is nonetheless recom-mended as a good practice. Again, you should be vigilant about checking with the vendor regarding upgrades to firmware and software because they may overcome some of the WEP problems.

As a general rule, 40-bit keys are inadequate for any system. It is generally accepted that key sizes should be greater than 80 bits in length. The longer the key, the less likely a comprise is possible from a brute-force attack.

WEP weaknesses

Security researchers have discovered security problems that let malicious users compromise the security of WLANs. These include passive attacks to decrypt traffic based on statistical analysis, active attacks to inject new traffic from unauthorized mobile stations (that is, based on known plaintext), active attacks to decrypt traffic (that is, based on tricking the access point), and dictionary-building attacks. The dictionary-building attack is possible after analyzing enough traffic on a busy network. However, the biggest problem with WEP is when the installer does not enable it. Bad security is generally better than no security.

When they do use WEP, they forget to periodically change static keys. Having many clients in a wireless network potentially sharing the identical key for long periods of time is a well-known security vulnerability. This is in part due to the lack of any key management provisions in the WEP protocol. When someone loses a laptop (whether lost or stolen), the key could become compromised along with all the other computers sharing that key. Shared keys can compromise a system. As the number of people sharing the key grows, the security risks also grow. A fundamental tenet of cryptography is that the security of a system is largely dependent on the secrecy of the keys. Expose the keys, and you expose the text.

Moreover, when every station uses the same key, an eavesdropper has ready access to a large amount of traffic for analytic attacks.

The IV in WEP is a 24-bit field sent in the cleartext portion of a message. This 24-bit string, used to initialize the key stream generated by the RC4 algorithm, is a relatively small field when used for cryptographic purposes. It is also static. Reuse of the same IV produces identical key streams for the protection of data, and the short IV guarantees that they will repeat after a relatively short time (between 5 and 7 hours) on a busy network. Moreover, the 802.11 standard does not specify how the IVs are set or changed, and individual wireless adapters from the same vendor may all generate the same IV sequences, or some wireless adapters may possibly use a constant IV. As a result, hackers can record network traffic, determine the key stream, and use it to decrypt the ciphertext.

The IV is a part of the RC4 encryption key. The fact that an eavesdropper knows 24 bits of every packet key, combined with a weakness in the RC4 key schedule, leads to a successful analytic attack that recovers the key, after intercepting and analyzing only a relatively small amount of traffic. This attack is publicly available as an attack script and open source code.

WEP provides no cryptographic integrity protection. However, the 802.11 MAC protocol uses a non-cryptographic Cyclic Redundancy Check (CRC) to check the integrity of packets, and acknowledge packets with the correct checksum. The combination of non-cryptographic checksums with stream ciphers is dangerous and often introduces vulnerabilities, as is the case for WEP. There is an active attack that permits the attacker to decrypt any packet by systematically modifying the packet and CRC sending it to the AP and noting whether the packet is acknowledged. These kinds of attacks are often subtle, and it is now considered risky to design encryption protocols that do not include cryptographic integrity protection because of the possibility of interactions with other protocol levels that can give away information about ciphertext.

Note that only one of the problems listed above depends on a weakness in the cryptographic algorithm. Therefore, these problems would not be improved by substituting a stronger stream cipher. For example, the third problem listed above is a consequence of a weakness in the implementation of the RC4 stream cipher that is exposed by a poorly designed protocol.

One of the flaws in the implementation of the RC4 cipher in WEP is the fact that the 802.11 protocol does not specify how to generate IVs. Remember that IVs are the 24-bit values that are prepended to the secret key and used in the RC4 cipher. The IV is transmitted in plaintext. The reason we have IVs is to ensure that the value used as a seed for the RC4 PRNG is always different. RC4 is quite clear in its requirement that you should never, ever reuse a secret key. The problem with WEP is that there is no guidance on how to implement IVs. The key, whether it is 64 or 128 bits, is a combination of a shared secret and the IV. The IV is a 24-bit binary number. Do we choose IV values randomly? Do we start at 0 and increment by 1? Do we start at 16,777,215 and decrement by 1? Most implementations of WEP initialize hardware using an IV of 0 and increment by 1 for each packet sent. Since every packet requires a unique seed for RC4, you can see that at volumes, the entire 24-bit space can be used up in a matter of hours. Therefore, we are forced to repeat IVs and violate RC4's cardinal rule of never repeating keys. Statistical analysis shows all possible IVs (2^{24}) exhausted in about 5 hours. Therefore, the IV is re-initialized starting at 0 every 5 hours.

Attacking WEP

There are several active and passive attacks for WEP, as follows:

- ✔ Active attacks to inject traffic based on known plaintext
- ✔ Active attacks to decrypt traffic based on tricking access point
- ✔ Dictionary-based attacks after gathering enough traffic
- ✔ Passive attacks to decrypt traffic using statistical analysis

Active traffic injection

Suppose that an attacker discovers the exact plaintext version of one encrypted message using a passive technique. The attacker can use this information to construct and insert correctly encrypted packets for the network. To do this, the attacker constructs a new message calculating CRC-32 values and performs bit-flips on the original message to encrypt plaintext to encrypted form. The attacker can now send the packet undetected to the access point. There are several variations of this technique:

- ✔ Destumbler (http://sourceforge.net/projects/destumbler)
- ✔ WEPWedgie (http://sourceforge.net/projects/wepwedgie)

Active attack from both sides

The attacker may make guesses on packet header contents rather than packet payload. Bit-flipping can transform destination addresses and route traffic to rogue devices where retransmission (with alterations) can occur. Educated guessing can also provide port information to allow passage through firewalls by changing it to use Port 80 (Web use).

Table-based attack

A small space of possible initialization vectors (IV) allows attackers to build decryption tables. Using passive techniques, the attacker gains some plaintext information. The attacker can then compute the RC4 key stream used by the IV. Over time, repetitive techniques allow an attacker to build a complete decryption table of all possible IVs. This allows an attacker to decipher every packet sent.

Passive attack decryption

IP traffic is redundant in nature and replication of this process easily yields enough data to decipher the encrypted text.

Monitoring

Monitoring is more of an intrusion than an attack, but it leads to further exploits. An attacker will monitor traffic until an IV collision occurs. A collision is when the algorithm reuses an IV. When a collision happens, the shared secret and the repeated IV result in a key stream that has been used before. Since the algorithm sends the IV in ciphertext, an attacker keeping track of all the traffic can identify when collisions occur. Then the attacker will use the resulting XOR information to infer data about the message content.

You can find commercial off-the-shelf (COTS) hardware readily available to monitor 2.4 GHz transmissions. By reconfiguring drivers, you can cause the hardware to intercept encrypted traffic. Using the techniques described previously, the WLAN becomes vulnerable.

Key management problems

WEP uses symmetric keys. This means that the algorithm uses the same secret key for encryption and decryption and that the sender and the receiver must possess the same key. Ah, the nub of the problem. There is nothing in the 802.11 standard about managing keys. Key management (probably the most critical aspect of a cryptographic system) for 802.11 is left largely as an intellectual exercise for the users of the 802.11 network. As a result, many vulnerabilities are introduced into the WLAN environment. These vulnerabilities include WEP keys that are non-unique, never changing, factory defaults or weak keys (all zeros, all ones, based on easily guessed passwords, or other similar trivial patterns). Additionally, because key management was not part of the original 802.11 specification, with the key distribution unresolved, WEP-secured WLANs do not scale well.

If an enterprise recognizes the need to change keys often and to make them random, the task is formidable in a large WLAN environment. When you have five laptops, this is an annoyance. When you have 5,000 workstations, this is a potential showstopper. Each one of those 5,000 workstations must have the same secret key, and the owner of every workstation must keep it secret. Generating, distributing, loading, and managing keys for an environment of this size is a significant challenge. Compromise one client and you have all the keys. You know what they say about secrets? Have you ever lost a laptop? Have you ever lost an employee? In both cases, you should change all 5,000 keys. Otherwise, someone can decrypt every message because everybody is using the same key. How often do you really think administrators will change the keys?

Protecting WEP Keys

One of the fundamental flaws of WEP is that it uses keys for more than one purpose. Generally, you don't use the same keys for authentication and encryption or the same key for integrity and privacy. Because WEP breaks these rules and others, it behooves you to protect your keys, since WEP doesn't provide any help here.

Default WEP keys

The manufacturer may provide one or more keys to enable shared-key authentication between the device trying to gain access to the network and the AP. Using a default shared-key setting is a security vulnerability because many vendors use identical shared keys in their factory settings. A malicious cracker may know the default shared key and use it to gain access to the network. Changing the default shared-key setting to another key will mitigate the risk. For example, the shared key could be changed to "95461" instead of using a factory default shared key of "11111."

NetGear Access Point uses the following four WEP sequences as default keys:

    10 11 12 13 14

    21 22 23 24 25

    31 32 33 34 35

    41 42 43 44 45

It is not surprising that a vendor has such simple default keys. What is surprising is that the first key didn't start at 11! In the event you don't know the default keys (well you do now) or you don't know whether there is a default key, check out www.cirt.net. Don't use default WEP keys!

No matter what your security level, your organization should change the shared key from the default setting because it is easily exploited. In general, organizations should opt for the longest key lengths (for example, 104 bits). Finally, a generally accepted principle for proper key management is to change cryptographic keys often and when there are personnel changes. Does your organization do this? Perhaps when you have 4 employees, but unlikely when you have 4,000!

The previous example showed we could use four different static keys. An access point transmits using only the first key, but can receive traffic encrypted with any of the four keys. Suppose that you have 100 users. Split them into four groups with four keys. This way, if any key is compromised, you need to change keys on only 25 stations, not all 100.

You can also use the third key as a key for the client to use to encrypt frames. The AP will use key 1 and the client, key 3.

It is worthy to note that some vendors generate keys after a keystroke from a user, which, when done properly, using the proper random processes, can result in a strong WEP key. Other vendors, however, have based WEP keys on passwords chosen by users; this typically reduces the effective key size.

You may find that your configuration utility doesn't have a password generator, but allows you to enter the key as alphanumeric characters (that is, a to z, A to Z, and 0 to 9) rather than as a hexadecimal number. Sounds like a good idea until you study it. Each character you enter represents 8 bits, so you can type 5 characters for a 40-bit code and 13 characters for a 104-bit code. Entering 5 characters in ASCII is not as strong as generating the key randomly in hexadecimal. Think of all the poor five letter passwords you could create. Another thing, an uppercase *A* is a different ASCII code than lowercase *a*.

Unfortunately, the IEEE 802.11 specification does not identify any means for key management (life cycle handling of cryptographic keys and related material). Therefore, generating, distributing, storing, loading, escrowing, archiving, auditing, and destroying the material is left to those deploying WLANs.

You just read a lot about the weaknesses of WEP. Table 11-3 is a summary of some of the more glaring weaknesses of WEP.

Table 11-3	WEP Weaknesses
Reference Number	*Weaknesses*
1	The IV value is too short and not protected from reuse.
2	The way keys are constructed from the IV makes it susceptible to weak key attacks.
3	There is no effective detection of message tampering (message integrity).
4	It directly uses the master key and has no built-in provision to update the keys.
5	There is no provision against message replay.

At a minimum, enterprises should employ the built-in WEP encryption. You're probably wondering at this point why the developers of the 802.11 standard chose RC4 for WEP. RC4 provides the following benefits for small organizations:

- The algorithm with a strong key (128 bits) and a sufficient IV (48 bits) is robust enough to protect data.
- The algorithm withstood attacks until recently.
- The algorithm is relatively efficient and uses fewer clock cycles than other algorithms providing comparable protection.
- It is an interim solution until AES replaces it.
- The patent owner, RSA, charges a small fee for the algorithm.

You can use WEP; however, we highly recommend 802.1X, WPA, AES, and proprietary technologies for enterprise WLANs.

Using WPA

You may have heard of 802.11i. If you haven't, check out Appendix B. IEEE 802.11i defines the robust security network (RSN). An access point will only allow RSN-capable devices to connect. RSN is the environment we are evolving to. It provides the security services we require for a network. Only time will tell whether there are flaws in 802.11i. We will cover 802.11i features in this section and later in the chapter when we cover AES. Implementing 802.11i will require new hardware. Not everyone will want or need to acquire new hardware, but will still want improved security. WPA comes to the rescue.

An initiative for improving WLAN security is the interim solution — Wi-Fi Protected Access (WPA) — to address the problems of WEP. WPA uses the Temporal Key Integrity Protocol (TKIP) to address the problems without requiring hardware changes — that is, requiring only changes to firmware and software drivers. TKIP is also part of the RSN.

WPA is an example of a software or firmware patch. The developers of Wi-Fi Protected Access originally called it WEP2. The joke around the Wi-Fi Alliance was something like, "When you build a new ship, you don't name it Titanic 2." As an interim security solution, WPA does not require a hardware upgrade to your existing 802.11 equipment, whereas the full-blown 802.11i does. WPA is not a perfect solution but is an attempt to quickly and proactively deliver enhanced protection to address some of the problems with WEP prior to the availability of 802.11i security features. It has two key features:

  ✔ 802.1X support

  ✔ Temporal Key Integrity Protocol (TKIP)

WPA uses 802.1X port access control to distribute per-session keys. Some vendors previously offered 802.1X support even though it was not specified in the standard. The 802.1X port-based access control provides a framework to allow the use of robust upper-layer authentication protocols. We cover this later in the chapter.

Temporal Key Integrity Protocol (TKIP) provides key mixing and a longer initialization vector. It also provides a Message Integrity Check (MIC) that prevents wireless data from being modified in transit. TKIP manages keys to prevent static key reuse. It also facilitates the use of session keys, since cryptographic keys should change often. TKIP includes four new algorithms to enhance the security of 802.11. TKIP extends the IV space, allows for per-packet key construction, provides cryptographic integrity, and provides key derivation and distribution. TKIP, through these algorithms, provides protection against various security attacks discussed earlier, including replay attacks and attacks on data integrity. Additionally, it addresses the critical need to change keys. Again, the objective of WPA was to bring a standards-based security solution to the marketplace to replace WEP until the availability of the full-blown IEEE 802.11i Robust Security Network (RSN), an amendment to the existing wireless LAN standard. RSN will also include the Advanced Encryption Standard (AES) for confidentiality and integrity.

Table 11-4 lists TKIP enhancements and demonstrates the WEP weaknesses it addresses. The numbers in the Addresses column refer to the numbered weaknesses (Reference Number) in Table 11-3.

Table 11-4	TKIP Enhancements	
Purpose	*Change*	*Addresses*
Message integrity	A message integrity protocol to prevent tampering	3
IV selection and use	A change in the selection of IV values and the reuse of the IV as a replay counter	1 and 3
Per-packet key mixing	A different encryption key for every frame	1, 2, and 4
IV size	An increase in the size of the IV to avoid IV reuse	1 and 4
Key management	A mechanism to distribute and change the broadcast	44 keys

AES-CCMP

WPA is still based on the RC4 algorithm, a stream cipher. But a major component of RSN is the use of the Advanced Encryption Standard (AES) for both data confidentiality and integrity. Presently, you can find AES WRAP (Wireless Robust Authenticated Protocol) products, but the final specification specifies the AES-CCMP (Counter Mode-Cipher Block Chaining MAC Protocol) algorithm.

The 802.11i specification offers AES-based data-link level cryptographic services that are validated under FIPS 140-2. Since AES will mitigate most concerns you may have about wireless eavesdropping or active wireless attacks, we strongly recommended its use. However, it must be recognized that a data-link level wireless protocol protects only the wireless subnetwork. Where traffic traverses other network segments — either local or wide area networks, including wired segments, the Internet, or your backbone — you also may require higher-level, FIPS-validated, end-to-end cryptographic protection.

The AES-based solution will provide a highly robust solution for the future but will require new hardware and protocol changes. Your organization may have difficulty justifying the use of AES as it will require you to build a Public Key Infrastructure (PKI).

Using Port Authentication

WPA and RSN provide port-based network access control. The Extensible Authentication Protocol (EAP) is a port-based authentication protocol that supports multiple authentication mechanisms (for example, tokens, smart cards, and digital certificates). The EAP specification doesn't care what authentication mechanism you choose to use, whether it includes the use of usernames and passwords, smart cards, biometrics, or PKI, or a combination of solutions (for example, smart cards with PKI). However, to be effective, your authentication solution must provide a reliable way of permitting only authorized users to access your network.

EAP (illustrated in Figure 11-2) is a standard, multi-vendor framework for combining port-level access control with authentication. The protocol defines messages exchanged between stations (supplicants), APs (authenticators), and back-end authentication systems. The mechanism blocks everything but EAP messages until the authentication server accepts the supplicant's access request.

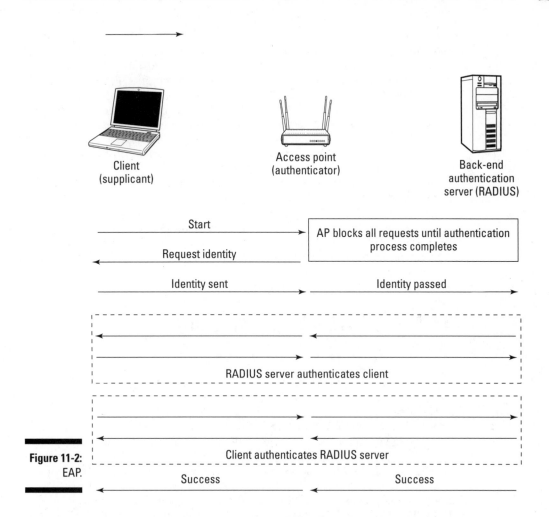

Client
(supplicant)

Access point
(authenticator)

Back-end
authentication
server (RADIUS)

Start

AP blocks all requests until authentication
process completes

Request identity

Identity sent

Identity passed

RADIUS server authenticates client

Client authenticates RADIUS server

Success

Success

Figure 11-2:
EAP.

EAP supports mutual authentication, key management, and dictionary-attack resistance. In addition, 802.11i defines the hierarchy for use with the TKIP and AES ciphers and a four-way key management handshake used to ensure that the station is authenticated to the AP and a back-end authentication server, when present.

You can implement IEEE 802.1x entirely on the AP (by providing support for one or more EAP methods within the AP), or you can utilize a back-end authentication server. The IEEE 802.1x standard supports authentication protocols such as RADIUS, Diameter, and Kerberos. You can use EAP for one-way or two-way authentication. The standard does not specify the authentication mechanism.

Typically, EAP runs over the link layer without requiring IP. It was originally used for Point-to-Point (PPP) remote access but is now being used by wireless network applications. Windows XP and many hardware vendors are building 802.1x security standards into their access points. For Windows 2000 Server, Microsoft implemented EAP in its Internet Authentication Service (IAS). Also, Cisco combined EAP with RADIUS in their LEAP security protocols for recent models of wireless access points, network cards, and CiscoSecure ACS. This provides a higher level of security than the typical WEP security. The 802.1x standard has a key management protocol built into its specification, which provides keys automatically. Keys can also be changed rapidly at set intervals. Check to see whether your access point supports 802.1x.

Security researchers have noted some security flaws in the 802.1x standard. This points out the need for good VPN technology despite this new standard. You can find an outline of 802.1x security issues at `www.cs.umd.edu/~waa/1x.pdf`.

Using LEAP, PEAP, and other forms of EAP

Your organization can implement the 802.1x standard with different EAP types, including EAP-MD5 (defined in RFC 2284 and supporting only one-way authentication without key exchange) for Ethernet LANs, and EAP-TLS (defined in RFC 2716, supporting fast reconnect, mutual authentication, and key management via certificate authentication). Currently, a new generation of EAP methods is being developed within the IETF, focusing on addressing wireless authentication and key management issues. These methods support additional security features, such as cryptographic protection of the EAP conversation, identity protection, secure cipher algorithm negotiation, and tunneling of other EAP methods. For the latest developments on the status of each specification, refer to the IEEE 802.11 standards Web site (`standards.ieee.org/getieee802/802.11.html`).

Like much in networking, the problem is not that there aren't enough standards, but too many standards. Following is a list of some of the more popular variants of EAP:

✔ **Lightweight EAP (LEAP)** (`www.cisco.com`): Mutual password authentication, challenge/response not encrypted, do off-line dictionary attacks. It is a Cisco proprietary protocol. LEAP dump for Red Hat Linux and Asleap (`http://asleap.sourceforge.net`). Cisco wants people to start using EAP-FAST.

- **EAP-FAST (**`www.ietf.org/internet-drafts/draft-cam-winget-eap-fast-00.txt`**):** Flexible Authentication via Secure Tunneling. Creates a tunneled authentication process. The tunnel establishment relies on a Protected Access Credential (PAC) provisioned and is managed by an authentication, authorization, and accounting (AAA) server.

- **EAP-TLS (**`www.microsoft.com` **or** `www.freebsd.org` **or** `www.linux.org`**):** Mutual certificate authentication, eavesdropping protection through the use of TLS. This is preferable, especially when running Win32 and already using certificates.

- **EAP-TTLS (**`www.funk.com` **or** `www.mtghouse.com`**) and Protected EAP (PEAP) (**`www.microsoft.com`**):** Authenticate servers by certificates and stations by passwords. Also tunneled over TLS. Works with Active Directory and NetWare Directory Service. Can trick into sending identity or credentials without protection of TLS tunnel; can intercept.

- **EAP-Subscriber Identity Module (SIM):** Uses Subscriber Identity Module (SIM) of a wireless handset. It has a possible use for roaming from WLANs to WWANs.

- **EAP-SRP:** Secure Remote Password; secure password-based authentication and key-exchange protocol. It provides good security but is not widely supported.

- **EAP-MD5:** Duplicates CHAP password protection on a WLAN. Earliest type; base-level. Not recommended for security-conscious enterprises.

Look at the list and you can probably pick the potential winner. Microsoft supports EAP-TLS and Cisco supports EAP-FAST. Which one supports the most widely-deployed operating systems?

EAP Questions

When looking for EAP products, you'll want to determine whether the proposed solution

- Provides adequate credential security
- Permits mutual authentication of the client and the network
- Supports or requires dynamic encryption keys
- Supports re-keying periodically
- Provides easy setup and management
- Fits easily into your network

When relying on usernames and passwords for authentication, it is important to have policies specifying minimum password length, required password characters, and password expiration. Smart cards, biometrics, and PKI have their own individual requirements and also require policy development.

Well, that is it for present and future network security features. Your organization may find that it is necessary to employ higher level cryptographic protocols and applications such as the point-to-point tunneling protocol (PPTP), layer 2 tunneling protocol (L2TP), secure shell (SSH), Transport-Level Security (TLS), or Internet Protocol Security (IPSec) to protect your information.

Chapter 12

Secure Wireless Access with Virtual Private Networking

*I*n Chapter 11, we show you ways to use encryption and other techniques to protect access to your network, helping ensure that only authorized users get connected. In this chapter, you discover ways to protect your data as it crosses someone else's wireless network. This way, using those hotel and airport wireless networks is safer, and you ensure that prying eyes are unable to see those cute pictures of your new niece.

Using a virtual private network (VPN) involves some work and isn't for the fainthearted, but we guide you through the process step-by-step, helping ensure that everyone benefits from this technology.

Secure Access with a VPN

Chapter 9 explores the risks of wireless networks. The possibility of eavesdropping is one of the risks. If you value the information on your network, it needs protecting. A VPN does that for you.

This chapter shows you numerous types of VPNs and allows for all levels of budget and skill. Hey, we know not everyone is rich and can afford those large commercial implementations.

Using a wireless network is a liberating thing. You continue working while getting a coffee, while at lunch, or even in the loo! What a picture that evokes. Ugh. Don't use your wireless connection in the loo; there are some places that should remain sacrosanct from such goings on.

All this additional freedom is good, but you don't want just anyone to peer at your innermost thoughts as your work travels across the wireless network, or especially across a public wireless network. It is a good thing to have unfettered access while traveling through airport lounges, coffee shops, and hotels, but only if your data remains private.

This is where a virtual private network helps. There are many types of implementations to use, ranging from commercial applications to those that are available as part of operating systems. Which one is best? Well, the answer is easy. It is the one you use!

Don't think that you can't afford to use a VPN. While commercial VPNs might be expensive, there is nothing wrong with using the myriad of other ones, especially in a small business in which the number of wireless users is small or the need is not as critical.

These methods also work on a wired LAN, so they serve a useful dual purpose if you connect from home over a DSL or cable network.

A VPN provides a special, secret tunnel between two networks that only authorized persons can access. This means you can use someone else's network, such as a hotel network, to get your company e-mail, while ensuring that no one else can see what you are doing. It's sort of like having a secret handshake, like all those secret organizations you hear about. Only by knowing the secret handshake do you get into the club. In our case, having access to the secret tunnel allows you to talk to the other members of our secret club.

The secret tunnel is set up and torn down each time you use it, similar to sharing the secret handshake each time you meet a new member of the club. No remnants remain. Gone. Sounds like a lot of work doesn't it? Well, believe it or not, after it is implemented and you have the client software needed on your workstation, all you do is run a little program and log on each time you need to connect to the office.

Defining the VPN

Webopedia defines *VPN* as an abbreviation of *virtual private network,* a network that is constructed by using public wires to connect nodes. For example, there are a number of systems that enable you to create networks using the Internet as the medium for transporting data. These systems use encryption and other security mechanisms to ensure that only authorized users can access the network and that the data cannot be intercepted. Whew! What does that all mean?

If you look at the words *virtual private network,* you can see that it has something to do with virtual, something related to private, and finally, something to do with a network.

- ✔ **Virtual** means existing or resulting in essence or effect, though not in actual fact. We can think of it, in other words, as *not real.* We are not creating an actual private network, but a pretend one on an existing network.

- ✔ **Private** indicates just that. It belongs to the parties involved and no one else. So you can carry on a conversation without anyone listening in.

- ✔ Finally, all this happens on an existing **network,** in our case a wireless network.

So why do you care? Using a VPN can allow you to access your mother's favorite recipes or that girlfriend's e-mail without worrying about anyone seeing them. If Colonel Sanders was using e-mail those many years ago across a wireless network, the secret ingredient to his chicken probably wouldn't have remained secret for long!

To implement a VPN, the first step is to determine where each endpoint will reside. This means that the VPN software and hardware need to be placed somewhere on your network, and client software to access it needs to be installed on your workstation or laptop. This is a one-way VPN. You connect to the VPN server and authenticate yourself in order to be allowed on the network to get your e-mail or whatever. There is another type of VPN that is two-way and is typically installed using encrypting routers, for example; however, this type of VPN is beyond the scope of this book.

Using the VPN technology allows you to connect to your office using a public network, such as the Internet, while ensuring that your communications are secure. A VPN connection across the Internet logically operates as a wide area network (WAN) link between you and the office. It gets a bit complicated doesn't it?

Try thinking of it this way. If you want to connect to your office and receive your e-mail or send a file to your colleague, you need to use some form of network, right? Perhaps you do this from home using a cable modem and your friendly local cable service provider. When you connect, your data travels across the cable provider's network until it reaches your company. Because your company does not have a direct link to the cable company, an intermediary was used — typically the Internet. So now you have three networks involved: your cable company, the Internet, and your organization. All of these can view the packets of data as they travel back and forth across these networks.

The one method used to reduce that potential viewing is to use yet another network, your virtual private network. This network runs on top of the others. So now you are using four networks to get the work done. This last one though, is what protects all those juicy e-mails from being intercepted by prying eyes. Figure 12-1 shows how all these networks connect. The wavy line represents your VPN protecting your data as it travels across the cable network, the Internet, and into your organization's network. What a great way to send Sally that recipe for fried chicken and know it's safe from prying eyes as it travels from your kitchen to your organization.

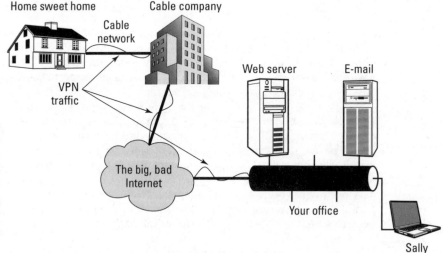

Figure 12-1:
Using four networks to connect to e-mail from home.

There are three general types of VPN. The more common one perhaps is the Remote Access VPN, allowing a mobile worker to securely access internal applications, such as e-mail or corporate applications, while on the road.

Next is the Extranet VPN that allows one organization to securely access another organization, typically over a leased line, although with more and more organizations opting to use the Internet instead. Finally, there is the Intranet VPN, in which data crossing the organization's normal network is encrypted. This last method is what we use to protect our wireless network.

In this chapter, we discuss the first and third of these methods.

VPN considerations

Considering whether to use a VPN entails deciding whether your information needs safekeeping. This one is pretty well a slam-dunk, as they say in basketball. Even normal business e-mail needs protection from prying eyes. Remember, we are talking about the world potentially seeing those messages. This is even more important if you are doing work on your wireless network that you consider confidential or sensitive.

Remember that even accessing your e-mail from the park outside your office is risky without a VPN because the person sitting near you may be tapping into your network and intercepting all your data. So you need to install and use a VPN to protect yourself, and by now are probably wondering how you figure out what to use.

A VPN is even more necessary when trying to connect to your organization to get e-mail or perform work while on the road. Rather than making a long distance (or 1-800) call to a corporate remote access number, the user calls a local ISP. Using the connection to the local ISP, the VPN software creates a virtual private network between the dial-up user and the corporate VPN server across the Internet.

There are numerous solutions available, from those that your MIS department purchases, installs, and maintains, to using an outsourced company to do the work for you. Of these solutions, there are commercial products and free products. Which method you choose depends upon a number of considerations.

A VPN implementation considers the following:

- ✔ **Address management.** The VPN assigns your network address on the private network and ensures that it is kept confidential.
- ✔ **Identification and authentication.** Your VPN verifies your identity and restricts VPN access to authorized users only. It also provides audit logs to show who is accessing what information and when.

- ✔ **Encryption.** Your data remains unreadable to unauthorized clients on the network.

- ✔ **Key management.** The product generates and refreshes encryption keys for the client and the server.

- ✔ **Multiprotocol support.** This need depends upon your plans for the virtual network. Typically, solutions handle the common protocols such as IP and Internet Packet Exchange (IPX).

- ✔ **Throughput.** Throughput depends on your particular needs. Collecting e-mail or sending the occasional file requires fairly limited bandwidth, especially if there are only a few users. Multiplying the number of users into dozens or hundreds and allowing Internet browsing along with other business-related activities requires faster, easily managed solutions.

- ✔ **Cost.** Cost-effectiveness remains crucial for any organization. This cost is driven by the number of users who connect and the degree of support needed.

The solution you decide on needs to reflect these items and fulfill your business needs. For example, I connect to my e-mail server using PPTP or SSH as I travel around the world. I use two methods primarily for backup purposes. Should one fail, I can still access my e-mail using the other method.

As a small business owner in the technology field, I have no real need for strong, more costly third-party solutions like a Cisco VPN solution or even GoToMyPC (https://www.gotomypc.com), a reasonably priced online alternative. I am able to support the VPN by myself with occasional help from my business associates. You may need a more robust solution, like one from Cisco, that includes technical support and specialized hardware. It depends on your level of expertise, the number of users you support, the criticality of getting access, and finally, your comfort level with the overall cost.

Understanding tunneling

Okay, so we are discussing this thing called tunneling, but what does it mean? Are we going to dig our way into a secure world? No, *tunneling* is a technique used by virtual private networks to hide traffic.

A tunnel is created using an accepted technique between two endpoints. Any data traveling between those two points is secured using encryption. Barry's tunnel has one endpoint on his laptop and the other on one of his internal servers in the home office. It is encrypted using an industry standard encryption protocol. It is important to note that you can have a tunnel that does not provide encryption. These are used merely to facilitate moving traffic from one protocol across a network that doesn't understand that protocol. For example, you can tunnel Novell's IPX protocol and send it across the Internet by tunneling it in the Internet Protocol (IP). However, any data he transmits from his laptop across the Internet while he is connected to his tunnel is safe until it reaches my inside network. After it's in my home office network, the data reverts to traveling unprotected over my local area network.

The data you are transferring (or *payload,* as it is called in the industry) become packets inside another protocol. Instead of sending a packet normally from your workstation or laptop, the tunneling protocol encapsulates the packet in an additional header. This additional header provides routing information so that the encapsulated payload can travel across the unsafe network.

These new encapsulated packets are then routed across the network. The logical path that the packets are using is called a *tunnel.* When the encapsulated packets reach their destination, the frame is unencapsulated and forwarded to its final destination. Tunneling includes this entire process (encapsulation, transmission, and unencapsulation of packets). In Figure 12-1, we show the number of networks you might use when implementing a VPN. In Figure 12-2, we show a diagram that highlights the tunnel aspects.

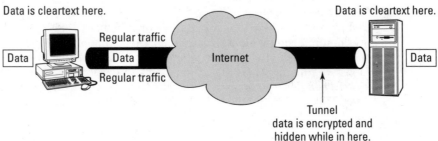

Figure 12-2: Tunneling traffic across a public network.

Data is cleartext here.

Data is cleartext here.

Regular traffic

Data

Data

Internet

Data

Regular traffic

Tunnel
data is encrypted and hidden while in here.

You can see from Figure 12-2 that your data gets carried inside a tunnel, shielding it from view. Remember that this tunnel gets created and broken down each time you connect and authenticate. It isn't sitting there waiting for you to use it and it is only useful to you, it is not shared. Each user creates his own tunnel when he connects, so, at any given time, there are numerous tunnels as your workmates connect from their hotel rooms while you all attend that conference boondoggle.

Tunneling technology is typically based on either layer 2 or layer 3 tunneling protocols. Layers are terms that correspond to the Open Systems Interconnection (OSI) Reference Model. Layer 2 protocols relate to the data-link layer and use *frames* as their unit of exchange. Point-to-Point Tunneling Protocol (PPTP) and Layer 2 Tunneling Protocol (L2TP), oddly enough, are layer 2 tunneling protocols. Layer 3 corresponds to the OSI Network layer, and layer 3 protocols use *packets*. IP Security (IPSec) is an example of a layer 3 tunneling protocol and many of the Cisco products use IPSec.

Whew! My head's spinning from all these technical terms. All this information is great, but the bottom line is finding a solution that works and is cost-effective enough for your needs. You do not need to understand the entire inner goings on as long as you use a commercial product and implement it as directed. We show you how to use a tunnel later in this chapter.

Finally, a word on *split tunneling*. This technique allows all the data going to your organization to be encrypted, but allows other data, such as your Internet browser, to remain unencrypted. This is a function of the VPN software you use, such as SSH. This allows the protection of sensitive or confidential data while allowing you to surf the Net using the hotel's or other service provider's network. This has the benefit of reducing network traffic on your VPN and letting you do two things at once.

There is a downside, however, and many organizations do not allow split tunneling. If someone breaks into your computer via the Internet while you surf the Net unprotected (you do use up-to-date virus protection all the time, right?), they could gain access to the corporate network. They might do this by compromising your machine, giving them administrator rights, and then gaining access to the desktop where your e-mail or other corporate applications are running. Many of the commercial VPN products allow the organization to add desktop policies to restrict what you do while connected to the VPN. This is a good thing to do to reduce any risk.

Deploying VPNs in WLANs

By now you have a basic understanding of what VPNs are and how tunneling works. Let's consider using these inside a wireless network. Many wireless access points have built-in support for remote access using virtual private networking. A remote user can direct a request to connect to the access point using the appropriate VPN port. If you are on a hotel's wireless network and provide the proper address, port, and associated VPN methodology, and then authenticate properly to the inside machine that is answering your request to connect, you will be connected and allowed on the corporate network.

But what we are looking for is protecting the wireless network itself with a VPN. One way to do this is to place all secured wireless access points in the organization on a single subnet. (You could set up a Virtual Local Area Network (VLAN) for this, or just use a subnet not in use within your organization.) Then ensure that the only way off this network and onto your corporate LAN is through a VPN server that requires authentication before allowing access. Once a user authenticates, all her traffic on the wireless network is encrypted and safe from prying eyes.

In this way, even if a hacker managed to penetrate your wireless access point's security and join the network, he would see no cleartext traffic because all legitimate users would be using an encrypted tunnel.

Wireless VLANs

A Virtual Local Area Network, or VLAN, is a network that is logically rather than physically segmented. It enables workstations and servers to associate independent of their physical attachment to the network. This is another example of a network running on top of another network, similar to the VPN we discuss previously in this chapter. So why do we do all this convoluted stuff to our networks? Good question. We are glad you asked.

One good reason for a VLAN is security. Utilizing this technique allows you to segregate your users more effectively, giving them access to only the resources they need. Because it is software-based, you can modify the design as needed without making physical changes to the underlying network architecture. Without a VLAN, making a network change involves modifying the physical topology and disrupting the users on that part of the network.

Okay, so there isn't really a physical topology when dealing with wireless networks, is there? You are right. But security is still important, isn't it? In fact, a VLAN is a great idea for your wireless network because it segregates the traffic and parses your users into groups, reducing the risk of accidental or intentional data interception.

One use for such a configuration in your wireless environment is to segregate user traffic from guest traffic. Suppose that you allow guests to your organization to access the wireless network in order to get e-mail or browse the

Internet while on your premises. You don't want them to see the general business traffic you might have on your wireless network, so you set up a separate network that only maps or identifies the e-mail and Internet gateways. Another use is to reduce the risk of handheld computers (or PDAs) that support only 40-/128-bit static-WEP co-existing with other wireless devices using stronger 802.1x with dynamic WEP. Placing the less secure devices on a dedicated network segment reduces your overall risk.

You need specialized equipment that supports VLAN technology in order to implement wireless VLANs. There are many vendor products, including Cisco's Aironet 1200 and Symbol's WS 5000 Wireless Switch.

To implement this functionality, you need both the specialized wireless equipment and the expertise to put it all together. Showing how it is done is not within the scope of this book, however.

Various Other Methods for Secure Access

You can use many different implementations of virtual private networking. These range from commercial third-party applications to those that are imbedded in operating systems. In this section, we describe how some of these operating system applications work. Since these are free (with your purchased operating system), they provide a ready VPN for many small businesses.

Using Microsoft's Point-to-Point Tunneling Protocol

Arguably the weakest of all the techniques, Point-to-Point Tunneling Protocol (PPTP) offers a quick and relatively painless method of access to your network. Although PPTP is not as strong a technique as the others, it offers a level of encryption that is more than adequate for most small-business owners. Better than that, unlike L2TP and IPSec, you do not need a certificate server and can implement it using native Windows commands. Finally, client software is available for all Microsoft operating systems and is supported by most commercial VPN vendors.

Microsoft built four phases of negotiation into each PPP dial-up session. Each one must complete successfully before the connection is established. First, the physical connection must be made. PPP uses Link Control Protocol (LCP) to establish, maintain, and end this initial connection. During its establishment, the authentication protocols are selected, but not actually implemented until later. In addition, a decision is made as to whether negotiation will use compression and/or encryption. The actual choice of compression and encryption algorithms and other details occurs during the last phase.

Next, the user's credentials are sent to the remote access server. Most PPP implementations use one of the four following methods to secure the password.

- **Password Authentication Protocol (PAP)** is a simple, cleartext authentication scheme. The password is in cleartext (unencrypted) and therefore is not one you use.

- **Challenge-Handshake Authentication Protocol (CHAP)** doesn't transmit the actual password. It uses a challenge, which consists of a session ID and an arbitrary challenge string, that's sent to the requesting client. The remote client then uses a special algorithm to return the username and an encryption of the challenge, session ID, and the client's password. The username is the only part sent in cleartext.

- **Microsoft Challenge-Handshake Authentication Protocol (MS-CHAP)** is very similar to CHAP. It provides an additional level of security because it allows the server to store hashed passwords instead of cleartext passwords. It allows a few other items, such as a password expired code and the ability to permit users to change their passwords. Version 2 (MS-CHAP v2) provides stronger security than CHAP for the exchange of the username and password credentials and for the determination of encryption keys. We won't bore you with the details. Take our word for it.

- Finally, to recap, it's in this step that the authentication data is collected and validated against the VPN user database or Windows domain controller, or other authentication server such as a Remote Authentication Dial-in User Service (RADIUS) server.

In the third step, there is an optional callback control phase. This uses the Callback Control Protocol (CBCP) immediately after the authentication phase. If you configure the session for callback, both the remote client and server disconnect after initial authentication. The server then calls the remote client back. This provides an additional level of security but also an additional level of administration because the callback numbers need to be maintained.

Finally, PPTP sets up the various network control protocols (NCPs) initally selected in Phase 1 to finalize all the parts needed to get connected. This is when you can access your e-mail or other network resources.

Using Windows XP for our client machine and accessing a Windows 2003 server, we show you how to install and use a PPTP tunnel for remote access in the following steps. (These steps are similar to those you can find in Windows 2000. So if you are using that operating system, follow along, noting any minor changes you may need.)

1. **From the Start menu, select Network Connections. The Network Selections screen appears. Click Create a New Connection.**

   You see the Welcome to the New Connection Wizard dialog box.

2. **Click Next. Choose Connect to the Network at My Workplace. Then click Next.**

3. **Choose Virtual Private Network Connection. Click Next.**

   You are asked to type in a name for the connection.

4. **Enter a name that will easily identify your VPN connection.**

   You see that Barry chose **My VPN Server** for this step, shown in Figure 12-3. Click Next to continue after filling in the name you chose.

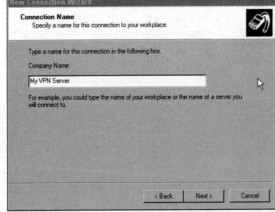

Figure 12-3: Entering a name for your connection.

5. **Windows will dial your ISP for you prior to connecting to the VPN. If you prefer to manually dial your ISP and then connect to your VPN server, select the checkbox marked Do Not Dial the Initial Connection.**

   We prefer to leave it on manual so that we can use any network connection as the conduit, rather than just the dial-up session.

6. Click Next to continue.

You now have to supply the IP address of the machine that will get you to your VPN software, as you can see in Figure 12-4. This is your wireless access point's IP address. Click Next after typing in the address, and you see the final dialog box.

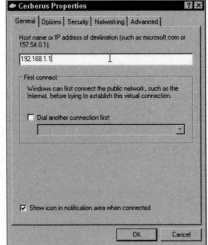

Figure 12-4:
Supplying the IP address of your VPN server.

7. You should select the check box next to Add a Shortcut to This Connection on My Desktop so that the VPN client is easily accessible. Choose Finish to compete the task.

8. The Wizard will then ask if you want to connect to your dialup service. For now, click No.

There are a few other steps to complete before testing the connection. An icon is added to your desktop.

Your access point now needs to be configured to allow PPTP access. This step depends on the model of your wireless access point, so you may need to consult the manual.

9. Connect to your wireless device in Administrator mode.

Typically this is achieved by pointing your browser at the IP address of the wireless access point. You see a login screen that varies depending upon the device. It asks for the administrator account and the password you used when setting up the wireless access point (see Chapter 5).

PPTP uses port 1723. Depending on your access point, you may need to supply this number. On most access points, it is an option you set, and is dependent on the many access points in use today. On one of our routers, this setting appears under Advanced Options, as shown in Figure 12-5. The Private IP address is the machine running PPTP inside our network that our client software connects to when we obtain access using PPTP. There are additional options you can use to restrict the time of day you want to allow PPTP access.

We don't use the time-of-day restriction option as we travel around the world. The different time zones make it difficult to figure out when we could or could not connect. It is useful, however, for those who travel in one or two time zones who want to reduce the potential for unauthorized access by disallowing access for those times of day when users aren't connecting through PPTP.

On a D-Link *Air*Premier AG access point, the option is found under Misc⇨Tools⇨VPN Passthrough. You need to consult the manual that came with your access point to determine the proper method for setting up PPTP.

Figure 12-5:
Using the menu to configure your access point to allow PPTP.

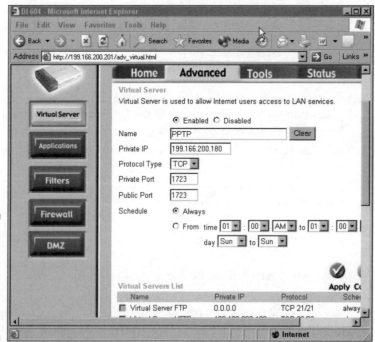

After you have set up your access point for PPTP, you can test it by running the client on your desktop from the icon created in Step 7.

10. Select and double-click the icon.

You should see something like Figure 12-6. The password field is hidden, as it should be, with the field allowing you to change it.

Leaving the password field completed is the easiest way to allow unauthorized access to your network. All that encryption is for naught if someone steals your machine. They will be able to connect automatically by double-clicking this icon. Make sure the Save This Username and Password box is not selected. This forces you to type them in each time.

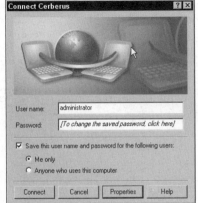

Figure 12-6:
Logging on using your VPN client.

In Figure 12-7, you see one of our PPTP connectors. If you click Properties, you see the IP address you entered under the General tab. Clicking Cancel returns you to the main window.

11. Click the Connect tab after entering your Username and Password.

The request is sent across the network to your organization. Once authenticated, a new network connection is added, and you can access the resources of your organization.

12. Remember to click the network icon when you are done and click Disconnect to terminate your network connection.

Following each of these steps enables you to connect to your network using the PPTP protocol that is free with Windows. It provides a reasonable level of security for small businesses and is simple to implement.

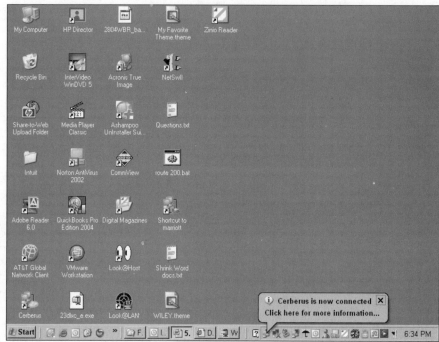

Figure 12-7:
A PPTP
desktop
icon and the
connected
network.

The security of the tunnel is very good; however, the weak link is your use of a password rather than a certificate. This means the degree of security depends entirely on your password generation skills. Choose a difficult password — after all, this tunnel is allowing access to your internal network. Do not use Granny's name or the pet dog or your own name with some numbers. Those are just too easy to break. Because the password travels across the encrypted tunnel, it is not accessible to password sniffing, only brute-force attacks.

Turn on logging in your Windows environment and view the logs each day to be sure no one is trying to guess your passwords. Also be aware that this method does leave you vulnerable to a denial-of-service attack if you use the same user account and password for both the tunnel and your day-to-day work. If you can attempt to gain access to the tunnel, so can anyone else; it's an open invitation. The strength in protection comes from protecting the passwords. The ideal method for protecting passwords is to get rid of them and use challenge-response or smart card technology. Unfortunately, this is less typical than it should be because of complexity and cost issues.

Considering Layer 2 Tunneling Protocol

Microsoft's implementations of Layer 2 Tunneling Protocol (L2TP) and Internet Protocol Security (IPSec) are available on the Windows 2000 and 2003 platforms and are designed to provide the highest possible security. Unfortunately, as a result of this increased level of security, these VPN solutions require the deployment of a Public Key Infrastructure, along with Pentium-class processors.

A Microsoft L2TP/IPSec VPN Client is available that allows computers running Windows 98, Windows Me, and even legacy Windows NT Workstation 4.0 to use L2TP connections with IPSec. I doubt very much if anyone still uses these clients; they are so old. However, should you be one of those, I have three words for you: Get new clients. Easy, eh? Really, neither of the Windows 98/Me clients offers security, and NT is no longer supported. You should be moving up to XP by now for the added support, security, and total cost of operation benefits available.

L2TP allows IP, IPX, or NetBEUI traffic to be encrypted, as we mentioned earlier, and then sent over any of the various network types, such as IP (the most obvious for us), X.25, Frame Relay, or ATM.

L2TP uses IPSec to start encryption earlier than the PPTP connection, providing greater security. It also allows for stronger encryption algorithms to protect the data.

Finally, IPSec provides data integrity, which proves that the data isn't modified in transit; replay protection, which prevents anyone from resending a captured packet stream; and data confidentiality by using encryption. PPTP only provides data confidentiality.

As we mention earlier, this is a more robust, secure method but needs more work to implement it. Perhaps in another book.

Using Windows IPSec

IPSec is an industry standard for encryption that Microsoft includes in its newer Windows 2000, XP, and 2003 operating systems. It is reasonably easy to set up between Windows machines and offers excellent security. Its primary weakness for the small business owner is its need for a certificate server or third-party certificate to ensure encryption. As you already noticed, it is typically used in conjunction with the L2TP protocol.

IPSec has two encryption modes: *tunnel* and *transport.* Tunnel encrypts the header and the payload of each packet, while transport only encrypts the payload. On your inside network, transport is sufficient because you are less concerned about anyone knowing your network topology, since they are likely authorized users who know the IP address ranges anyway.

One reason for using transport mode inside the network is the small gain in encryption speed; however, for a small network, the overall cost in speed of encrypting traffic is minimal. One Microsoft expert we talked to said it costs about 1 to 1.5 percent of the network bandwidth to use an IPSec tunnel. That is a pittance compared to the overall enhancement in security that is gained. Remote access, however, definitely requires tunnel mode to hide those inside IP addresses from prying eyes.

IPSec is a good method of protecting your wireless network if all your client workstations and servers are Windows 2000, XP, or 2003. After setup, no one is able to see any of the traffic between machines unless they have the proper credentials. However, IPSec tunnels only support IP traffic and therefore cannot be used for IPX or other network traffic.

While IPSec is not overly difficult to set up, it is beyond what we can provide in this book. Use the Microsoft Web site and download one of their excellent *Step-by-Step* series of articles, one of which guides you through using IPSec between Windows machines.

Oldies but goodies — SSH2

SSH is an abbreviation that stands for *Secure Shell,* which is a program allowing you to secure network services running over an insecure network, such as the Internet. This is another tunnel, similar in idea to the ones we have discussed throughout this chapter.

The Secure Shell concept originated on Unix and therefore has been around the block, so to speak. Its origin goes back to the early days of Unix and a need to protect the weak services that Unix had implemented. Today, it is commonly used to tunnel services with cleartext passwords such as Telnet and FTP. These dinosaurs are widely used and effective, hence their popularity, but they use cleartext passwords! It boggles the mind that in today's computing world, so many organizations think so little of security that they still use such tools.

The current version of SSH is version 2 (hence the *SSH2* in a title of this section). Discover details about it in the IETF-secsh Internet-Drafts on the site `www.ietf.org/ID.html`. There is more available information than you ever wanted if you search through all the various drafts.

SSH also allows you to securely log in to remote host computers, just like we do using PPTP. This allows you to run commands on a remote machine, providing secure encrypted and authenticated communications between two machines or networks. Within this tunnel, you run the services you want to protect, such as e-mail, FTP, or even Web browsing. Barry tunnels a number of items, including his e-mail, Web browsing, and even a Terminal Services connection to an inside server.

To use SSH, you need to operate the server portion of the program on a machine inside your network. You then use the client to connect to this server software and establish a tunnel. SSH server is free within the Unix world and is often installed by default, making it kind of easy to use. But as time and Windows advanced across the world, the need for a Windows version of SSH became evident, and that need was fulfilled, allowing you to use this oldie but goodie even in that competing product line. The downside, of course, is that the SSH Server for Windows doesn't come free, costing around a thousand dollars.

The good news for the financially strapped is the possibility of using a free Unix SSH server and letting your Windows clients connect to that. Client software that recognizes either Unix or Windows servers is available for all the major workstation operating systems.

SSH provides mutual authentication as the client authenticates the server, and the server authenticates the client. This way, both parties can be sure they are dealing with the correct party. Each party uses either certificates or public keys to ensure the identity of the other.

As we mention earlier in this chapter, Barry uses two remote access methods. This is his second method for getting into his home office network. He set this up a number of years ago with the able assistance of a good friend so that they can use these tunnels to connect to the outside world while on assignment with various clients.

Finally, one of the really good things about SSH is its ability to use public key cryptography or certificates. This is far stronger than a mere password. There is a great deal of good information at `www.ssh.com`. We recommend visiting the site to learn more about SSH, including the steps needed to implement and support it.

Who Is Doing the Talking?

We all recognize that some communications require confidentiality, integrity, and authentication — the foundations of security. The adoption of cryptographic techniques or, more commonly, encryption and the keys used within that, provides the degree of security needed. Putting such encryption into place, along with the ongoing management of the keys and algorithms, needs an infrastructure. This infrastructure is commonly referred to as a *Public-Key Infrastructure,* or PKI.

On the plus side, using a PKI immensely enhances your security and allows you unbridled freedom to perform business over any network. On the downside, putting this structure into place and then managing the day-to-day operation of it is expensive and requires considerable technical expertise.

This PKI methodology is being adapted and optimized to fit the wireless world's (WPKI) needs. It consists of the same components that are used in a traditional PKI. These include an End-Entity (EE), the Registration Authority (RA), the Certification Authority (CA), and the PKI directory. In addition, a new component referred to as the PKI Portal is required.

Remember, you can think of a PKI as being the components that allow you to use certificates and encryption along with all the parts you need to put them together and manage them. However, few organizations today are using a PKI mainly due to the complexity and cost, along with different competing standards that make sharing a PKI between business partners difficult.

Simply put, the steps involved in using such a mechanism after it is installed include the user's End-Entity software requesting a certificate from the PKI Portal, which forwards the request to a Certification Server. The Certificate Server issues the certificate and posts it in a directory for later use. The portal sends the location of the certificate back to the End-Entity that requested it. Content servers use the directory to retrieve the certificate and its revocation dates for use in authenticating the user. The user device then uses that certificate to issue secure requests to applications, such as Web portals, and the data flows in an encrypted form between the user device and the application, ensuring that no one sees or tampers with the information.

This is all great stuff isn't it? However, this short explanation doesn't really touch on the complexities involved in implementing a Wireless PKI (or any PKI for that matter). It might highlight for you, though, that such technology is available and, should your business have such a need, you can implement fully secure methods of accessing your applications across a hostile, open network such as the Internet.

Part IV

Keeping Your Network on the Air — Administration and Troubleshooting

In this part . . .

After you plan, set up, connect, and secure your wireless network, you must manage that network and keep it up and on the air. Troubleshooting a wireless network involves far different issues than troubleshooting a wired network, including Fresnel zones, free space loss, and contention issues. Luckily, this part provides direction on those issues as well as providing you with sound advice on expanding the distance of your network using bridging techniques. You see how to perform traffic management and learn how to monitor for performance issues and trouble spots. Finally, in this part, you see how to find all your access points and detect and respond to intrusion.

Chapter 13

Problems with Keeping on the Air

· ·

· ·

*T*his chapter helps set out processes and steps for managing that new wireless network and ensuring that it runs as trouble-free as possible. Like any network, implementing it is the first step, but living with the results and constantly tweaking the parameters to keep the network humming is another thing altogether. Sometimes it can be tough to be the network person. We help ease that burden by providing information on typical trouble spots and how you can prepare to overcome them.

Troubleshooting Redux

In Chapter 16, we discuss a number of tools and methods for helping run a wireless network; there, we also recommend annual audits to ensure that it remains functional and secure. Here we discuss an approach to troubleshooting to provide you with enough information to discover where problems are — and how they might be resolved.

We notice that true analytical troubleshooting capabilities seem hard to find. Folks know their products and equipment but are hard-pressed to take a step-by-step approach to analyzing the issue, research methods, or techniques to resolve the issue and implement the solution. Too often, we see network people misunderstand the actual issue and take inappropriate steps or place blame where it doesn't belong instead of attempting to solve the problem. We show you one way to bypass all that and actually fix the problem.

The following broadly defined steps are a good starting point:

1. *Know your network.* What does it consist of in terms of access points, users, LAN connections, and client devices?

2. *Determine the actual problem.* Much effort is wasted analyzing a problem that doesn't exist because someone used the effect instead of the cause as the base assumption.

3. *Get help early.* Don't waste time thinking that you can do it all. Know where your technical library is and who is strongest on each aspect of your network. A team is always better than one.

4. *Break the problem down into components and review each one.* Is the problem that users cannot connect? Then determine precisely where they cannot connect, when they cannot connect, how they are attempting to connect, and what exactly happens when they attempt to connect. Often, getting the exact information from the user rather than their translation of that evidence helps immensely.

5. *Determine which aspect of the network is failing.* Avoid using the effect that a user is experiencing; that can be misleading. Step through each component and ensure that it is functioning correctly until you reach the actual problem area. Although it may seem intuitive to just go right to the cause, you can often solve the problem faster by being rigorous in your approach.

6. *Fix one problem at a time.* Doing too much at once can hide the real solution. Try one thing at a time, noting what happens and whether it repairs the problem before trying the next thing.

7. *Don't automatically assume two things are broken at once.* Although this is possible, it's unlikely and only complicates your efforts.

8. *Isolate components where possible and see whether they work correctly before placing them back on the network.* However, don't just swap parts. This does nothing to increase your problem determination skills.

9. *After the issue is identified and repaired, test it.* Be sure it is working and that you know why it didn't.

10. *Document the issue, its cause and effect, and how it was resolved.* Building a troubleshooting document can pay dividends the next time something happens.

You can obtain oodles of information from the vendors of your products, including common troubleshooting steps and specific details on configuration errors. Use these resources.

Table 13-1 describes some common errors that occur.

Table 13-1	Common Configuration and Other Errors
Error	*What to Do*
Unplugged	You'd be amazed at how often a component is unplugged accidentally. Check it first.
Loose cable	Check all connections and ensure that they are tightly coupled.
Disconnected	Ping each component on the network and ensure that you can reach them.
Network card malfunctioning	Is the user's network card functioning correctly? Often, this is the problem and not the rest of the network. Verify that it is properly installed.
Incorrect SSID	Ensure that the user has the correct SSID or network name in her wireless network card.
Incorrect channel	Make sure that all devices are communicating on the correct channel. This is 1–11 for North America.
Incompatible standards	Are all the devices using compatible 802.11 standards? Remember that a client with an 802.11b network card will be unable to use an 802.11a access point.
Inaccurate WEP/WPA settings	Has the user inadvertently turned off WEP or keyed in the incorrect key? Is WPA configured accurately?
Network address incorrect	Is DHCP working correctly and assigning the correct IP addresses? Do an `ipconfig /all` command on Windows clients and ensure that the IP address information is correct.
Dual DHCP	Are multiple access points each using DHCP? If so, check for conflicts and set each one to supply only particular subnets.
MAC conflicts	Are you using MAC address security? If so, is the list of approved MAC addresses kept up-to-date and accurate?
Weak signal	Maybe the user in is a location not supported well by your wireless network. Verify the location against the site survey or use an analyzer to detect how strong the signal is and whether it will support connectivity.
Interference issues	Check the signal in the area for interference from newly installed refrigerators, microwaves, or other items that can impact a signal.

Any of these errors can severely impact your network. Of course, we haven't discussed all the other pieces, such as bridges, routers, and switches. If you follow the steps covered in this section, however, you should be well on your way to resolving any network issues that you encounter.

Am I in Your Fresnel Zone?

Are you a friend of Fresnel? First off, get the pronunciation correct. The *s* is silent — like *fren EL,* with apologies to dictionary lovers the world over. Fresnel is a type of focusing system made up of hundreds of prisms, which amplify and focus light into a narrow beam so that it can be seen miles away. It was discovered of course, by Augustin Jean Fresnel of France. In the wireless world, he provided the means to calculate how out of phase deflections between the transmission source and the receptor will be in a given situation. Why will they possibly be out of phase? Good question. Go to the head of the class.

There is no *s* sound when pronouncing Fresnel. Leaving it out will help let others understand that you know what you are talking about in the wireless world.

When you transmit your wireless radio waves, they generally spread out from your transmitter. As they spread out, they form an ellipsoid. Those signals that travel in the most direct line to the receiver form the best signal. Those that are spread out — and subsequently are deflected by objects, trees, buildings, and air currents — get progressively worse depending on the extent of their deflection.

If the spread-out waves don't bump into anything, they just travel off into the air until they disappear. However, if they bump into something (or get deflected), they may end up at the receiving antenna. If so, they will probably be *out of phase* with the straight-line signals and therefore have a *phase-canceling* effect, which reduces the power of the arriving signal. You can see an example Fresnel zone in Figure 13-1.

Water is arguably the most critical aspect. A building's walls allow the signal to pass reasonably freely, but objects containing water deflect easily. Trees, bushes, and people contain water, so keep them out of the Fresnel zone. Line of sight gives you only a part of the picture — you may set up your antennae in spring before the trees are full and think that because you can see the other antenna, it should be okay. It won't be. Not only will the branches block the signal, but transmission also worsens as the leaves develop.

Fresnel zone

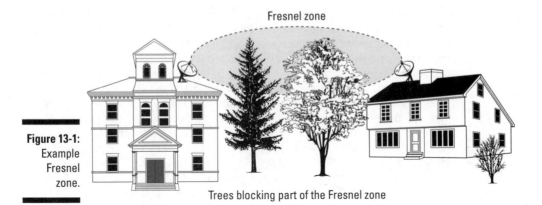

Trees blocking part of the Fresnel zone

Figure 13-1:
Example
Fresnel
zone.

We show you how to manually calculate the Fresnel zone in your network in Appendix C. To calculate your particular Fresnel zone, you can go online at www.zytrax.com/tech/wireless/calc.htm or many other locations and either use the calculator shown or in the case of this site, download the code and run it on your own machine. You see an example of the calculator from this site in Figure 13-2.

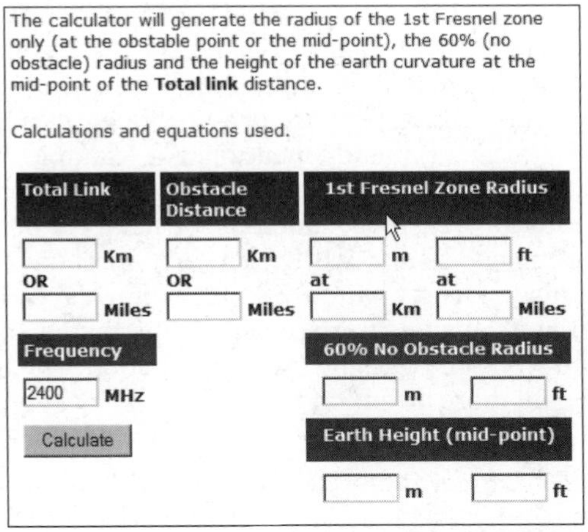

Figure 13-2:
Example
Fresnel zone
calculator.

Enter the distance between the antennae and then click the Calculate button. The Web page then shows you the radius of the first Fresnel zone along with Earth Height and Obstacle Radius.

What do all these things mean to you? If you have no external antennae, then it's probably just interesting reading. However, if you're trying to connect multiple locations based on some distance, you need decent line of sight (LOS) and a clear Fresnel zone. Even when you believe you have a clear LOS, you might not have a clear Fresnel zone because of those objects that we mention earlier. Your wireless signals propagate outwards and, of course, not all of them end up being direct to the other antenna. How much they are deflected ends up impacting your overall wireless signal. Because those signals not traveling directly to the receiver are being deflected, when they arrive at the other end, they create an out-of-phase condition and can result in canceling out the direct waves. If the distance is long enough (about 5 kilometers or more), even the curvature of the Earth can have an impact. That is one reason for using a calculator to determine these factors and then adjusting your signal accordingly.

How do you adjust the signal? One obvious method is to raise the antennae so that they are high enough to maintain a clear Fresnel zone. Another method is to relocate them to achieve the same effect. You might also change the type of antenna and use one more suited to your particular needs. A final option (that we would really disagree with it) is to cut down any trees that may interfere. Naturally, this should be a last resort and should be done in accordance with any laws in your neighborhood.

All these factors may impact your network and cause you to wring your hands over troubleshooting problems. In LOS networks, revisit these components and verify that nothing has changed. Remember that trees grow — and what worked last year may no longer work because of a now-taller tree. Also, trees in general are tricky objects, as we already mention. For example, in summer, they may cause errors in your network but give you no problems in winter. They may allow your radio waves one day and not the next. It's best to just avoid them altogether. You need to also verify that your antennae haven't become misaligned because of heavy winds or loose bolts. Maybe ice is covering them in the winter and causing interference. Maybe the Earth's bulge changed and now impacts your line of sight. Okay, not that one, we hope, or we will all be wishing we'd paid more attention in all those survival-type movies.

Multipath Interference

Perhaps your life is a crossroads, and you have many paths you can choose. Choose the wrong path, and life may not be as sweet as you'd like it to be. This is basically what happens with your wireless signals sometimes. It gets deflected on its journey, and that can cause problems.

Multipath propagation is what happens when a radio signal takes different paths when being sent out from a source (for example, your access point) to a destination node (for example, the other access point). As the signals travel toward the other antenna, items get in the way, like walls and doors and equipment, which causes the signal to bounce around in different directions. Some of the signal may go directly to the destination, and other parts may bounce from a desk to the ceiling and then on to the destination. As a result of all this, some of the signal encounters delay and thus travels a longer path to the receiving access point.

This delay causes the information in the 802.11 signal to overlap, which confuses the receiver. This is often referred to as *intersymbol interference* (ISI). If the delays are great enough, bit errors in the packet occur. The receiver can't distinguish the symbols and therefore interprets the corresponding bits incorrectly.

Multipath interference causes downfade, upfade, corruption, and nulling. The negative effects induced on a WLAN by reflected RF signals arriving at the receiver along with the main signal.

Delay spread is the difference in time between the main signal and secondary (reflected) signals arriving (< 4 nanoseconds). This results in

- Decreased signal amplitude (downfade)
- Corruption
- Nulling
- Increased signal amplitude (upfade)

Some multipath solutions include

- **Antenna diversity:** Antennae on single input
- **Switching diversity:** Antennae on multiple receivers
- **Antenna switching diversity:** Antennae on multiple inputs
- **Phase diversity:** Adjust phase of antenna to phase of signal
- **Transmission diversity (used by most WLAN manufacturers):** Transmits from antenna last used for reception

When this happens, the receiving station detects the errors through 802.11's error-checking process. The cyclic redundancy check (CRC) checksum that is always computed will not compute correctly, indicating that errors are in the packet. In response to those errors, the receiving station will not acknowledge the source, so eventually, it is retransmitted by the originator. If these retransmissions occur too often, they begin to degrade performance, and your users will be unhappy with their service levels.

This is more likely to occur in locations with lots of metal objects, such as factories and warehouses, than in regular office buildings. It is still something to keep in mind, though, because perhaps your office adjoins a factory and your signals would bounce on its equipment.

When comparing the different spectrums — frequency hopping spread spectrum (FHSS), direct sequence spread spectrum (DSSS), and orthogonal frequency division multiplexing (OFDM) — the most susceptible to multipath propagation is DSSS, which is the one used in 802.11b networks. FHSS uses relatively narrow channels (1 MHz) and changes transmit frequency often, making it difficult for multipath to occur. OFDM (used in the 802.11a and 802.11g spectrums) transmits information on a number of subchannels, which helps reduce the impacts of multipath for those spectrums. 802.11b systems then are usually the most susceptible, so moving off those onto the other types is a potential solution when you are not too heavily involved in 802.11b equipment.

Another solution may come from the vendors. Palo Alto, Calif.-based Airgo Networks (www.airgonetworks.com) recently unveiled its AGN100 Wi-Fi chipset, which it indicates will actually use multipath interference to its advantage. This chipset listens in all directions at the same time; by simultaneously processing all that information, you apparently get a strong signal. We will have to wait and see whether this works as advertised. It also has the disadvantage of needing to be incorporated into all the access points, thus making it problematic if other vendors don't buy into it.

You Can't Go That Far: Free Space Loss

Free space. (There isn't much that is free these days, is there?) Unfortunately, loss we can do without. Especially if you're in a casino reading this book in between card hands. *Free space loss* is the power loss of the radio wave traveling through the air with no obstacles impeding it. In other words, it's the distance it will travel if let be and nothing tries to impact it. Maybe *Star Trek* fans will think that means it will travel into outer space and other galaxies. We doubt it because the signal just isn't always that strong.

So how far is far? There really isn't such a thing as an unimpeded signal because something always gets in the way, whether a person, tree, building, weather, or whatever. It is primarily caused by *beam divergence,* which is the signal energy spreading over larger areas at increased distances from the source, much like the beam of a flashlight. There is, though, a correspondence between free space loss in dB and distance. You can find mathematical models to determine this in Appendix C.

The decibel (dB) is the basic unit of measurement used in Wi-Fi radio signals. The *B* is in honor of Alexander Graham Bell, who was the inventor responsible for much of today's acoustical devices.

The formula for this loss at 2.4 GHz is

```
FSL = 104.2 + 20 log D
```

where D = distance in miles.

Example: At 5 miles, FSL is 118 dB.

You can use the following guideline when calculating free space loss: When you double (or halve) the distance from the transmitter to the receiver, the signal level lowers (or increases) by 6 dB.

This loss is *attenuation,* which is simply a reduction of signal strength during transmission of a signal. The free space loss attenuation needs to be taken into consideration when designing your network to ensure that your signal reaches its intended antenna, especially when that distance is large.

As the frequency increases, so too does path loss, meaning that a 2.4 GHz system has a greater range than that of a 5 GHz system of equal power output because of its lower frequency. A 2.4 GHz radio signal typically experiences a free space path loss of about 120 dB over a distance of 5 miles. This isn't a problem for indoor setups but is problematic when you're planning a larger scale network.

To help counteract this loss, you need to either increase the sensitivity of your devices or boost the signal with repeaters. All this should be coupled with the data in your loss budget (see Chapter 2). When you design a network, you start with output power, add antenna gain, and then subtract loss from your cables and the free space loss. If the resulting number still exceeds the equipment's receiving sensitivity, the signal gets through. We recommend providing for a margin of error by defining a fade margin of perhaps 20 dB.

Contention-Free Frames

Collisions occur, whether on the highway or on a network. Managing those collisions is what differentiates the better network. To do that, you need to use some form of detection with enough smarts to keep the collisions to a minimum while ensuring that traffic actually passes across the medium in a timely manner.

The basic mechanism in use is *Distributed Coordination Function* (DCF). To use this mechanism, our wireless networks use Carrier Sense Multiple Access with Collision Avoidance (CSMA/CA) for managing potential frame collisions. Most LANs use a similar but different protocol called Carrier Sense Multiple Access with Collision Detection (CSMA/CD). Wireless cannot use the Collision Detection method for a couple of reasons, one of which is that the radios would have to transmit in Full Duplex, which is far more expensive, so they try and avoid the collision rather than detect it.

When you operate a wireless network, detecting collisions is hard, so CSMA/CA just tries to avoid them, effectively managing the problem. In CSMA/CA, the Medium Access Control (MAC) layer uses the *Distributed Coordination Function* (DCF) protocol that works as listen-before-you-talk scheme. Too bad more people don't use that, isn't it? Another factor is the *Point Coordination Function* (PCF), which is an optional function used to implement time-bounded services, like voice or video transmission. This Point Coordination Function makes use of the higher priority that the access point gains by using a smaller Inter Frame Space (PIFS). By using this higher priority access, the access point issues polling requests to the stations for data transmission, thereby controlling network access. In order to allow regular stations access to the network, each access point must leave enough time for Distributed Access in between the PCF. The following lists some of the key aspects of DCF and PCF.

   ✔ **Distributed Coordination Function**

   • All stations contend for access.

   • Available with BSS, ESS, and IBSS.

   • AP similar to wired hub; used to send data.

   ✔ **Point Coordination Function**

   • Contention-free frame transfers.

   • Requires an AP, so only BSS and ESS.

   • AP polls stations.

Along with that is the clear channel assessment (CCA) algorithm that measures the RF energy at the antenna and determines the strength of the received signal, which results in the measured signal Received Signal Strength Indication (RSSI). The protocol has a threshold rule for the RSSI signal strength; if the threshold is below a certain level, the MAC layer is given the clear channel status for data transmission. If it is above the threshold, no clearance is given for communication. In that case, the station waiting for clearance waits for a determined length of time and tries again. This timeframe is the DCF Interframe Space (DIFS) and is used to establish clearance to retransmit. The medium must remain idle for the DIFS time period or no clearance is given.

However, the station cannot remain idle forever, or it would never communicate. Thus, another option is available to allow the station to send frames, using Request to Send (RTS), Clear to Send (CTS), and acknowledge (ACK) transmission frames. The station begins by sending a short RTS frame. This includes the length of the message and the destination. Included is the network allocation vector (NAV). This NAV is used to alert all other nodes in the networks to wait for the duration of transmission. After seeing this NAV frame, the receiving station sends a Clear To Send frame, echoing the sender's address along with the NAV item. If the sender does not receive this CTS frame, it assumes that a collision occurred and sends another RTS frame, in effect starting over again. If the CTS frame is received, the transmission begins, starting with an ACK frame for verification. Between two consecutive frames in this whole sequence, a Short Interframe Space (SIFS; a sort of time-out period) gives the devices time to respond. These SIFS are shorter than the DIFS period, giving both the CTS responses and the ACKs the highest priority access across the network. This does, however, initiate a high level of overhead on the network. You can use the On with Threshold setting for large packets, though, which should help. Whew! That's quite a load, and you might want to take a minute to breathe again.

Collisions still occur, of course, but this hopefully minimizes the number of collisions, keeping the network running efficiently. Numerous technical manuals explain this in more detail, but we hope that this short summary provides a decent overview of the process. If you're hungry for more, try the book by Ramjee Prasad, Werner Mohr, and Walter Konhauser, *Third Generation the Mobile Communication Systems* (Artech House). You can also go to encyclopedia.thefreedictionary.com/CSMA-CA and read about it there. Another excellent article can be found at www.sss-mag.com/pdf/802_11tut.pdf.

Hidden Node — So Where Is It?

So now we have nodes that are hiding from us? Yikes! Do they have a life of their own? No, this is another technical aspect of wireless networking. This term refers to those nodes or stations that are out of range of the others; this often occurs with outdoor installations. Of course, it can also happen indoors, like when you have two workstations separated by an interior wall that causes the signals to break up, allowing them to hear the access point but not each other. If we use the example of a typical topology with an access point and a number of stations nodes surrounding it in a circular fashion, each station must be in communication range of the access point, or they cannot communicate. The stations, however, cannot always hear each other's traffic because of obstructions like trees or buildings.

These hidden stations can therefore disrupt network traffic by improperly sending at times when other nodes are transmitting. This results in interference and back-off behavior that reduces network performance. That's a bad thing. It's even more vicious when the network is using things like streaming video, causing performance to possibly drop by as much as 70 percent. The collision avoidance mechanisms discussed earlier just aren't effective in dealing with this problem because they were never designed to handle today's continuous data transmissions.

The RTS/CTS method discussed earlier was designed to resolve the hidden node problem although a paper exists that indicates it doesn't always fix the problem. A detailed technical discussion on this appears at `nislab.bu.edu/sc546/sc546Fall2002/blocknode` where the proponents outline the problem and possible solutions. This is a time- and bandwidth-consuming process that is required for every transmission by every wireless node. And apparently, it still doesn't address the problem because more than one node might initiate this process at the same time because they cannot hear each other directly.

So what other solutions exist? It seems that there are mixed messages depending on vendor implementations. The KarlNet company (`www.karlnet.com`) offers the TurboCell product, which uses a centralized control function at the access point or base station to help eliminate hidden stations. The TurboCell access point uses a specially optimized polling technique to tell the wireless stations when they can transmit. It uses this and a free-for-all technique that prioritizes the stations to avoid the issue. You'd have to try the product, we'd guess, to be sure it works for you.

You might also investigate the Wireless Central Coordinated Protocol (WiCCP), which purports to eliminate the hidden node problem. WiCCP is said to be a protocol booster for 802.11b wireless networks, providing cyclic token-passing medium access and also scheduled allocation of the available network resources to eliminate the hidden node problem. You can find out more at `www.patraswireless.net/software.html`. Better yet for those on a budget, it appears to be a freely available solution.

Finally, you can consider the following:

- Use RTS/CTS to reduce impact.
- Increase station power.
- Remove obstacles.
- Move stations.

You also need to consider the Near/Far condition and implement solutions to it. This occurs with

- ✔ Multiple clients nearer to AP with high power settings
- ✔ One or more client farther away with lower power setting

Some of the solutions for Near/Far conditions include

- ✔ Increase power to remote station.
- ✔ Decrease power to local stations.
- ✔ Move the remote station closer to the AP.

Managing Power

Ah, power. The aphrodisiac of many people. From politicians to business people to kings and queens. And now you can find it in wireless networks as well. But, of course, here we talk about power in the literal sense of electricity, not those other types of power.

We all know and love the need for power in our laptops and digital assistants. The more, the better, right? On long plane trips, a two-hour battery just isn't that effective anymore. It's one reason airlines are beginning to slowly add power outlets to their seats so you can use your device for as long as you desire. This is handy for us because we often need to finish that chapter or research the next one while winging merrily away to some foreign realm.

To read more about the realities of electrical power, you can peruse Appendix C, where we go into excruciating detail for you. Suffice it to know, however, that increasing the power to your access point might increase the signal strength and allow you to reach that far point.

The FCC allows only 4 watts of radiated power from an antenna in a point-to-multipoint wireless LAN connection using unlicensed 2.4 GHz spread-spectrum equipment, so beware of increasing past this amount.

Your access point will use a certain level of power, typically between 30–100 mW. Changing this increases the potential signal strength and may allow for that slight extra reach that you are looking for in your access point.

On certain Linksys equipment, you might use SNMP to change the power settings. You can go to www.pasadena.net/aprf for an interesting article on doing this yourself. There is a page at www.personaltelco.net/index.cgi/AccessPointReviews showing numerous access points and their power ratings that you can use for reference. Some even include whether the power level is changeable; however, remember that this will probably invalidate any warranty you might have with the device.

Power over Ethernet (PoE)

Power over Ethernet (PoE) is mentioned elsewhere in this book, but here we tell you more about what this is and how it works. Some access points can be powered by using the Ethernet cable that connects the access point to the wired network. This is typically implemented by using a specialized piece of equipment in your wiring closet that inputs AC power along with the data connector from the wired switch, and then outputs DC power over some of the unused wire pairs in the networking cable that runs between that special module and your access point. This eliminates any need to run a power cable to the access point, thus allowing more discretion where it is placed because there is no need for an outlet nearby.

This is an IEEE 802.3af PoE standard, so it stands up to some scrutiny. The IEEE began the process in 1999; early players included 3Com, Intel, PowerDsine, Nortel, Mitel, and National Semiconductor. It was formally approved by the IEEE Standards Board on June 12, 2003. Using such a mechanism allows for more freedom in selecting a location that best suits the radiated radio waves, allowing for optimal access point placement. This is especially useful in old buildings or locations where running electrical power might be problematic.

Two types of devices are specified in this standard: Power-Sourcing Equipment (PSE) and Powered Devices (PD). The PSE provides 48v (volt) DC power, with a current limit of 350 milliampere (mA), to the PD and is limited to a continuous maximum power output of 15.4 watt (W). Dual-radio wireless access points typically require around 14 W of power, so there is ample there.

In addition, there needs to be enough cumulative power available to support all your connected PoE devices. This cumulative power can quickly add up to a large amount, possibly more than what is being supplied by a standard 110v AC wall power switch. Large PoE installations therefore may need additional 110 or 220 AC power lines.

How does it all work then? Power passes from the Power Sourcing Equipment to your powered device over standard Ethernet CAT-5/6 cables. Ethernet signals travel along two twisted pairs, one pair for each direction. There are four twisted pairs in each CAT-5/6 cable. PoE uses one spare pair for the positive DC supply and the other spare pair for the negative return. Another method involves actually using a pair of wires that's already being used to pass data. Either implementation provides power to the device. You can see how it works in Figure 13-3.

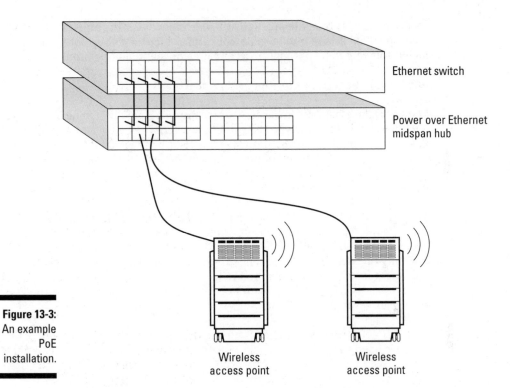

Ethernet switch

Power over Ethernet
midspan hub

Figure 13-3:
An example
PoE
installation.

Wireless
access point

Wireless
access point

This is really useful for only those sites that require it because it does add complexity — when you add complexity, things can break or be more difficult to manage. If you plan to purchase and use such a facility, be sure to visit www.poweroverethernet.com and read all about it in detail before jumping into something you may regret later.

Calculating your Power Budget

Your *Power Budget* is the total power output from your wireless system and is also the sum of the following items:

- ✔ Radio transmit power
- ✔ Cable and connector losses
- ✔ Antenna gain

Put these into a calculator (www.zytrax.com/tech/wireless/calc.htm) with the correct figures, and you have the budget for your installation. The fundamental aim of your radio link is to deliver sufficient signal power to the receiver in order to achieve a performance objective. This objective is typically specified as a minimum bit error rate. Configuring your power budget allows this rate to be optimal. You can find a hugely scientific primer on this at `wireless.ictp.trieste.it/school_2004/lectures/carlo/linkloss`, but beware that it is not for the faint of heart.

Managing the available power in your wiring closet might also be a challenge. Each port adds 15.4 watts to the total power required. This adds up quickly in larger installations. A small wiring closet supporting 100 PoE users needs to deliver up to 1,540 watts of power simultaneously to those users. Add this to the 1,000 watts that the average switch requires, and your 2,540 W switch will need 23-amp service.

That can easily be a problem for many wiring closets. In most buildings, electricity is rated for only 15- or 20-amp service. Many LAN switches support 240–360 ports in a single chassis, requiring 35- to 50-amp service to support PoE fully on each port. You will need a lot more power into the closet than you might have originally planned to use. You might want to make sure that if you're constructing a facility with PoE, multiple 20-amp circuits and/or 220v service are available.

Those of you with Cisco devices can use the Cisco Discovery Protocol (CDP), which is used to specify the exact power requirements of each device after initial power has been placed on the line. When using this proprietary method, Cisco says that it can power more devices with the same power budget, which might be a big help in those large installations.

The following lists some of the things to look for when purchasing PoE equipment:

✔ Do the Power Sourcing Equipment (PSE) and Powered Devices support power classification?

✔ What is the maximum wattage the system can deliver?

✔ How much power will the PSE draw from the wall?

✔ How does the PSE deal with a loss of power from a failed power supply?

✔ Will you need special power provisions in your wiring closet to support the PoE switch port density you require?

After you implement a Power over Ethernet infrastructure and are happy with it, you may want to think about power surge protection. The Citel Inc. company (`www.citelprotection.com`) has designed surge protection that prevents electrical damage over PoE installations. Its MJ8-505-24D3A60 product features a circuit that isolates data from Ethernet power lines, protecting all eight pins. The clamping response of this product apparently permanently eliminates transients, ensuring equipment safety and optimal data transfer.

Another vendor, smartBridges (`www.smartbridges.com/web/products/po.asp`), has introduced its PoE Outdoor product, which is a unique Power over Ethernet device for outdoor wireless installations. This has NEC-compliant surge protection and weatherproof durability, so it offers some pretty impressive-sounding capabilities.

On a smaller scale, the Injector product is a passive device that gates the power jack onto the Ethernet connection and is used to inject power onto the Ethernet cable. The box has two RJ-45 jacks: One has power, and the second has only Ethernet connections.

The Injector has voltage surge protection on the Ethernet-only side of the injector, and also surge protects the power lines, shunting any surges to the ground side of the power jack. It can handle an 800v surge and 1500 W burst, thus providing some degree of protection against these scourges of all power installations.

Many new product lines are being introduced with PoE capability. You'll need to contact your favorite vendor for the latest details.

Chapter 14

Bridging Networks to Manage Coverage

*B*eing able to network devices without the need for cabling is a major boon for small and medium businesses alike. Although large enterprises flood wire to make their office buildings network-ready, anyone setting up a small or medium business is unlikely to have the same luxury. For one thing, you may not want to go through the hassle of wiring up your office if you don't own the building — you may move on to new premises when the business takes off.

As a tenant, you may not have sufficient space on one floor or in one building to house all your employees. In the not-so-distant past, you had to call your Regional Bell Operating Company (RBOC) and get them to install a leased line to join your networks in the two buildings. Or you had to get the landlord to agree to let you run wire from one floor to another. This represented a large outlay and commitment of money for a small or start-up organization. You could use an on-demand service such as dial-up, but you won't get the through-put you need.

You are ready to extend your network beyond your indoor LANs. Perhaps you are fed up with the high recurring costs associated with leased lines or the expense that comes with running fibre underground, especially in areas where right-of-way issues exist. Now, with Wi-Fi and Wi-Fi5 standards, you have another option. You can use RF to move data from one wired segment to another using a wireless bridge. And since you are using unlicensed and readily available technology, it is affordable.

Using That Site Survey

In Chapter 2, we emphasize strongly the need for a site survey upfront before investing in any technology. Well, planning and implementing bridging and switching is where the site survey comes in real handy. If you did your work properly, you will know whether you have a viable link: You need RF line-of-sight for a viable link. When you are bridging two buildings, this is essential. When you are bridging indoors, this is not so much the case. Can the antenna at one end of the connection see the antenna at the other end? Will you need to install an antenna tower? Will you need to remove obstructions? Do you have a positive link budget? Is the signal strong enough? If you don't know the answers to these questions, you need to revisit your site survey. Under ideal conditions, 802.11 signals are capable of spanning distances of 25 miles or more, but who's working under ideal conditions?

Bridges and Switches and Routers, Oh My!

In earlier chapters, we discuss two different types of wireless networks: ad hoc and infrastructure. You see that ad hoc is another name for independent basic service set. We use the term *ad hoc* because wireless clients can spring up at any time and form a network with another wireless client. Similarly, we use the term *independent* to denote that these clients are not tied to a wired network; they are on their own.

This chapter focuses not on ad hoc networks, but on infrastructure mode. When you think of wired infrastructure, you most likely think of the wiring that snakes its way throughout your building. But there are other components, such as the patch panel, hubs, bridges, switches, and routers. The wireless network has an infrastructure, as well. We talk about one important component of our wireless infrastructure: the access point. By definition, every wireless access point (WAP) is a wireless-to-Ethernet bridge. However, in this chapter, we will switch gears and bridge the gap between the access points and other infrastructure devices. More specifically, we will focus on bridges and switches. These interconnectivity devices are elements of our wireless infrastructure.

In general, a network bridge provides a portal for frames of data to cross between segments or subnets. It is a Layer 2 bi-portal device connecting two segments or subnets. Any filtering we do is based on the Layer 2 header. You can use bridges to join segments on different floors in the same building or in different buildings. You can use bridges to join segments of different types: token ring to Ethernet or wireless to Ethernet. Basically, the bridge determines whether the frame should traverse the other segment, boosts the signal, and passes it on when applicable. Think of the bridge as an office partition between two parts of your office. The bridge looks at the MAC address and decides whether to throw the frame over the partition to the other side.

Switches are similar to but different from bridges. Bridges can do no more than simply pass frames based on source or destination MAC addresses. On the other hand, switches can send frames directly to the network segment of the addressee. The switch can act as a filter and can eliminate unwanted traffic. In the main, when you have a small network and you are not concerned with broadcasted data, you connect all your devices in a star configuration to a hub. The hub acts as a multi-port repeater by regenerating the signal and sending it to all attached devices. As your network grows, you can use a bridge to segment traffic and keep local traffic local. However, every time you add a segment to the bridge, the traffic on all segments increases because all devices on all segments broadcast their messages. This is the essence of a broadcast technology such as CSMA/CD and CSMA/CA. All devices on the same network will hear all broadcasts, since they are in the same broadcast domain. Switches, on the other hand, will only forward data addressed to a particular host on a particular segment. They make more efficient use of the available bandwidth.

Understanding Wireless Bridges

Connecting LANs wirelessly requires the use of wireless bridges. Until recently, wireless bridges were expensive and intended primarily for enterprise use. There are a few dedicated bridges, but generally you use an access point with bridging functionality. Some bridges use proprietary standards and won't interoperate with Wi-Fi equipment. However, most new products are Wi-Fi compliant. D-Link sells the DWL-G810 108 Mbps 802.11g Wireless Bridge (`www.dlink.com/products/?pid=241`), Linksys sells the WET11 Wireless Ethernet Bridge (`www.linksys.com/products/product.asp?grid=33&scid=36&prid=602`) and WET54G Wireless G Ethernet Bridge (`www.linksys.com/products/product.asp?grid=33&scid=36&prid=603`), and Cisco sells the Aironet 1400 Series Wireless Bridge (`www.cisco.com/en/US/products/hw/wireless/ps5279/products_data_sheet09186a008018495c.html`). If you want to use 802.11a, RadioLAN (`www.radiolan.com`) sells the BridgeLINK-11a 802.11a Wireless Bridge.

TIP

Note that your manufacturer may use different names to describe the following operating modes.

In essence, a wireless bridge provides connectivity between two wired LAN segments. You can configure the bridge as either point-to-point (P2P) or point-to-multipoint (P2MP). Point-to-point mode (also called master/slave or LAN-to-LAN) connects two LAN segments by using two bridge units. To connect two geographically separated LANs, you can use an access point as a point-to-point wireless bridge. Figure 14-1 illustrates the use of a wireless bridge in point-to-point mode. Point-to-multipoint mode lets you construct a network that has multiple bridges talking to each other wirelessly.

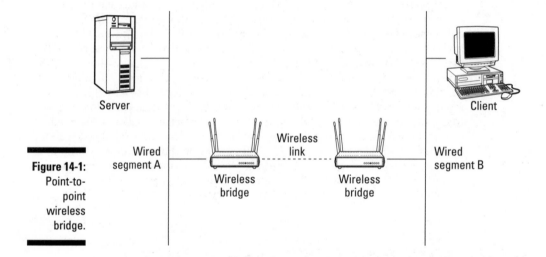

Figure 14-1:
Point-to-
point
wireless
bridge.

Most bridges have the same configuration and management options you'd find with an access point. This means you have a console port, telnet, HTTP, SNMP, or a client-side program.

A wireless bridge is a half-duplex device capable of Layer 2 wireless connectivity only. Wireless bridges have options similar to those of an access point. Chapter 5 explains these options. Some of the options you find on a bridge are as follows:

   ✔ **Fixed or detachable antenna.** Detachable antennae are a desirable feature on a bridge. With a detachable antenna, you can mount the bridge indoors and run a cable to the antenna mounted outside. If you don't have a detachable antenna, you most likely will need to mount the bridge outdoors in a NEMA-compliant weatherproof enclosure. You can find information about the National Electrical Manufacturers Association (NEMA) at www.nema.org. Among other things, NEMA specifies standards for enclosures to protect the contents from the effects of weather.

✔ **With or without antenna diversity.** As a rule, bridges don't need antenna diversity because they are static devices (unlike your clients). That is, the bridges sit at each end of the link, and the area around the bridge tends not to change too much.

✔ **Semi-directional or directional antenna.** When using a bridge in point-to-point configuration, you want to focus the beam as much as possible using a directional antenna. When using a point-to-multipoint configuration, you most likely would use a semi-directional antenna. You would not use an omni-directional antenna in either case. The signal radiating behind the antenna is wasted and may in fact be a security risk.

✔ **Filtering.** Like access points, bridges usually have MAC-address and protocol filtering built in. As the administrator, you can program the wireless bridge to allow or disallow network access to any device based on its hardware address. You can also filter traffic based on Layer 3 datagrams, Layer 4 datagrams or packets, and Layer 7 data.

✔ **Modular radio cards.** Some wireless bridges can act as an access point and bridge simultaneously. When using one radio card, throughput suffers. Optionally, you can add another radio and use different radios for access and bridging, which increases the throughput.

✔ **Variable output power.** Similar to the access point, you can control the radiation pattern of the bridge and limit its reach.

✔ **Power over Ethernet (PoE).** Many bridges support PoE when you just don't have a power outlet where you want to install the bridge.

✔ **Wire diversity.** Bridges support 10BaseTx, 10/100BaseTx, 100BaseTx, and 100BaseFx.

You should establish a full-duplex connection to the wired segment to maximize the bridge's throughput. Also, you should take note of how far it is to the wiring closet when specifying wired connectivity options for the bridge.

Wireless bridges working with other wireless bridges operate in one of the following four modes:

✔ Root mode

✔ Non-root mode

✔ Access point mode

✔ Repeater mode

Using root mode

When you have more than one bridge, one bridge must act as the root bridge. A root bridge can communicate only with non-root bridges and other client devices but cannot associate with another root bridge. Figure 14-2 shows a root bridge communicating with non-root bridges.

Using non-root mode

Bridges in non-root mode associate wirelessly with bridges in root mode (see Figure 14-2). Some wireless bridges support client connectivity to non-root mode bridges while the bridge is in access point mode. This mode is actually a special mode in which the bridge is acting as an access point and a bridge simultaneously. Client devices associate to access points (or bridges in access point mode), and bridges talk to other bridges.

You may find some vendors calling this offering a *Wireless Distribution System* (WDS). WDS is a bridging mode in which the access point can simultaneously bridge to another access point and act as an access point to clients.

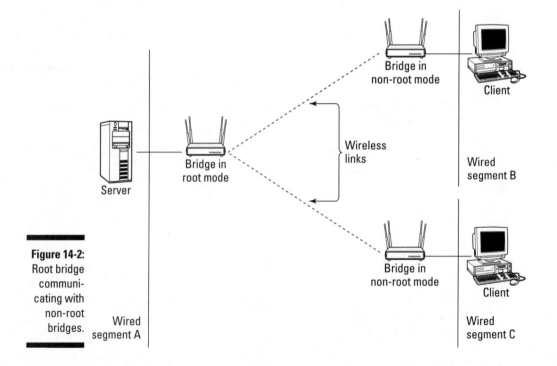

Figure 14-2:
Root bridge communicating with non-root bridges.

When using the Spanning Tree Protocol, all non-root bridges must have connectivity to the root bridge.

Using access point mode

Some manufacturers give you the option to connect clients to bridges, which is actually the same as giving the bridge access point functionality. In some cases, the bridge has an access point mode that converts the bridge into an access point exclusively. In this mode, a device set to master mode can communicate with slave units, as well as with wireless clients within its range. Slave units don't have the same ability and can communicate only with the master unit.

Using repeater mode

You can also define wireless bridges as repeaters. In repeater configuration, you place a bridge between two other bridges for the purpose of extending the length of the wireless bridged segment. Although using a wireless bridge in this configuration has the advantage of extending the link, it has the disadvantage of decreased throughput due to having to repeat all frames using the same half-duplex radio. Repeater bridges are non-root bridges, and many times the wired port is disabled while the bridge is in repeater mode. Figure 14-3 shows a bridge in repeater mode.

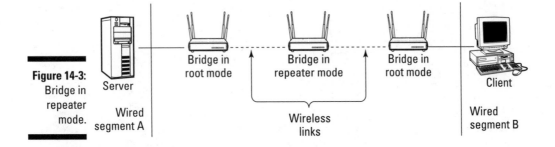

Figure 14-3:
Bridge in repeater mode.

Server

Bridge in root mode

Bridge in repeater mode

Bridge in root mode

Client

Wired segment A

Wireless links

Wired segment B

Wireless Workgroup Bridge

A device that is similar to a wireless bridge, and often confused with it, is the wireless workgroup bridge (WGB). The biggest difference between a bridge and a workgroup bridge is that the latter is a client device. When you look at the association table on an access point, you see the WGB as one device, not all the devices it represents. On the WAP, you don't see the MAC addresses of the devices connected to the workgroup bridge. The WGB is primarily used to connect one LAN (perhaps aggregating several computers networked via wired Ethernet) to another network by connecting wirelessly to a wireless access point. In a way, it is like a multiplexer. You can also use it as a wireless network device, connecting directly to another wireless-equipped computer in so-called ad hoc mode. Figure 14-4 shows you how to use a workgroup bridge on a network.

Clearly, you may not have a need for a wireless workgroup bridge in your home. But it may find a niche in your work environment. For instance, you may have a wired segment that is physically isolated (say, a portable classroom or office), but you want to connect everyone to the wired network.

You will find that wireless workgroup bridges have most of the options of a bridge discussed previously: fixed or detachable antennae, MAC and protocol filtering, variable power output, and varied wired support.

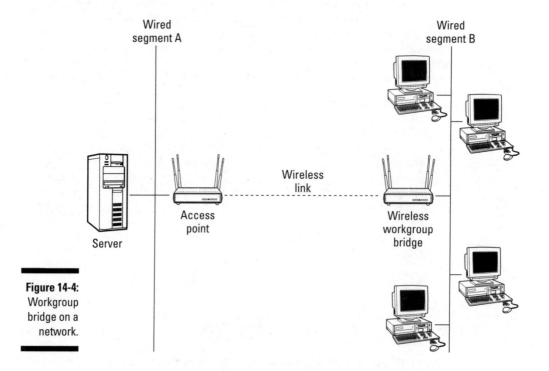

Wired segment A

Wired segment B

Wireless link

Server

Access point

Wireless workgroup bridge

Figure 14-4:
Workgroup bridge on a network.

When connecting wired users through a workgroup bridge, you should know that there generally is a limit to how many clients can live on the wired segment. The number of clients ranges from 8 to 128 depending on your manufacturer. Regardless of what your manufacturer says, you will find that when you have more than 30 clients, you will start to see degradation, and your clients might complain.

Going to the Movies: A Bridge Too Far?

Determining whether your wireless bridges can connect between two buildings is not as simple as determining whether a wireless client can connect to an access point. When locating outdoor bridges, you should consider the following:

- The physical location of the sites at each end of the link
- Line-of-sight between antennae
- The use of modern radio technology to reflect signals

Before you think you can just point your home access point at another access point to create a link, you should know that the path loss for one mile is approximately 104 dB (see Appendix C). On top of that, should you intend to send the signal any distance, you will need to worry about line-of-sight, Fresnel zone, and earth bulge. To recap, microwave signals travel in a straight line, but they spread as they travel out. The required beam clearance for the signal spread is known as the Fresnel zone, which is an imaginary ellipsoid surrounding the straight-line path between the antennae. Other chapters cover the first two concerns, but we have not talked about the earth's bulge. Even on terrain that seems flat, the earth curves. If any of you are Platygæanists, then forgive me for this next sentence. The earth is not flat, even though people still believe it is (reference the Flat Earth Society at `www.flat-earth.org`). By the time you travel out about seven miles, you must account for the earth's curvature when planning for longer paths. You can use one of the following topographical tools to do your planning:

- Plotpath (`www.v-soft.com/PlotPath/Index.html`)
- Terrain Navigator (`www.maptech.com/land/terrainnavigator/index.cfm`)
- Topo USA (`www.delorme.com/topousa/default.asp`)
- TOPO! (`http://maps.nationalgeographic.com/topo`)
- TopoZone (`www.topozone.com`)

In Appendix C, we tell you that you can have blockage in up to 60 percent of the Fresnel zone. So you need to calculate the Fresnel zone. You can find calculators to help you at www.firstmilewireless.com/calc_fresnel.html, www.wisp-router.com/calculators/fresnel.php, or http://pizon.org/tools/fresnel.php.

Building Hardware Bridges

When selecting a wireless bridge, you should consider the following factors:

- ✔ **Ease of installation, setup, and maintenance.** To simplify, look for manufacturers that sell kits.

- ✔ **Availability of technical support.** Access points are fairly trivial; bridges and switches are not.

- ✔ **Availability of management and configuration utilities.** You will want remote configuration (for example, SNMP support) and good diagnostic utilities.

- ✔ **IEEE 802.1d (spanning tree) support.** When you have more than one bridge in your network, you could have more than one path. Without the spanning tree protocol, you could create an endless loop in which frames just keep getting passed. This protocol helps prevent that phenomenon.

- ✔ **Range.** You should consider output power, antenna gain, and your link budget.

- ✔ **Security.** You should select a product that supports your security policy (for example, authentication, encryption, MAC and protocol filtering). See Chapter 10.

An attacker can compromise your network through eavesdropping on your path or attempting to access the remote administration facilities of the bridge. Attackers can also create rogue bridges. If you are bridging two networks outdoors, obviously someone can set up a rogue bridge in between your two endpoints. You need to check for this periodically.

- ✔ **Total cost of ownership.** You may have to lay out money for training, operation, and maintenance, so don't just look at the original outlay cost.

If you are brave, you can find some bridging kits at WirelessNetworkProducts.com (www.hdcom.com/bridgekits.html), and you can make a bridge yourself. They sell bridging bundles for P2P and P2MP networks up to 2.5 miles apart using 802.11b or g technology (up to 54 Mbps). A typical kit includes two radios, two 10-foot low-loss indoor cables, two lightning arrestors, two 20-foot low-loss outdoor cables, two 14-dBi high-gain antennae, and antenna-mounting hardware.

To set up hardware bridges, follow these steps:

1. **Turn bridging on for each device. Put the MAC address of each bridge into the other bridge.**

2. **Set the SSID to the same value in both bridges. Remember our earlier discussion for naming SSIDs. Don't call it Bridge!**

3. **Set both bridges to the same channel: any value from 1 to 11 when using 802.11b/g technology in North America.**

4. **Set the operation mode to point-to-point or wireless bridge mode.**

   Your manufacturer may not use any of these terms, so check the documentation.

5. **Configure any other vendor-unique value.**

You should be good to go. If not, read the troubleshooting section that follows.

Bridging the Gap: Getting Started

Before hauling your equipment to the roof, you may want to ensure that your configuration works. Set the configuration up in a lab or someplace with little interference. Try it out. Better to find out now while it's easy to do something about it.

After you shake down the hardware and know you can make a link, take it outside. Again, pick a location where you can easily test it. Try your parking lot or go to the local high school and try it out on the football field. Hopefully, this test works as well. If not, make the adjustments.

Next, align your antennae where you ultimately intend to place them. This is a two-person job because you need one person to adjust each antenna to obtain the best signal. When everything is working and you are sending frames, mount the bridge in a weatherproof container.

Don't mount the bridge in a plastic container. Plastic becomes brittle and cracks when exposed to sunlight.

Troubleshooting Your Bridged Network

If you don't get a link, ask yourself the following questions:

- ✔ **Is everything attached?** Whenever you call a vendor's customer support line, they ask you whether the machine is turned on. But, they bite their tongue. However, this is a good place to start. Ensure the antennae and power cables are fully attached.

- ✔ **Are your bridges using the same firmware?**

- ✔ **Is there an adequate signal?** You may want to test the signal strength using your network discovery utility, wireless packet analyzer, or spectrum analyzer.

- ✔ **Is there an RF line-of-sight?**

- ✔ **Are the bridges set to the same channel?**

- ✔ **Is any other access point in the vicinity using the same channel?**

- ✔ **Is bridge mode selected on both devices?**

- ✔ **Are the antennae aligned?**

- ✔ **Is the antenna the right type?** For example, you should use a highly directional antenna. You may get by with a semi-directional, but you probably can't use an omni-directional antenna.

- ✔ **Is the correct MAC address in each bridge?**

- ✔ **Is there only one DHCP server for both bridges and attached devices?** You must configure the segments as part of the same routable network. If you don't, you will find it doesn't work. Don't use two routers, but do use two hubs or switches.

- ✔ **If you are using static IP addresses, are the subnets correct?**

- ✔ **Are the same security features enabled on both bridges?** You can't have WEP on one and not the other.

- ✔ **If using encryption, are the same keys used in both devices?**

- ✔ **Is the bridge protected from the elements?**

Building Network Bridges in Windows XP

If hardware isn't your bag, you can build a software bridge. You can take advantage of the bridging capability built into Windows XP. Microsoft calls this Network Connections feature a *Network Bridge*. The Network Bridge allows a computer running XP with multiple network adapters to act as a bridge, connecting different local area network segments. The computer acting as the bridge must have Ethernet support and a wireless network adapter. You can get value from this feature not just when connecting two networks. For instance, you may find bridging useful for connecting a laptop with Centrino to a cabled network when you don't have a wireless access point — you just add a wireless card to any other computer on your network and it can bridge the wireless onto the cabled network.

Creating a network bridge

First, you wire all your computers together using a hub or inexpensive switch. Then, you install a wireless adapter in the computer used for bridging. After you install the network interface cards and connect them, go to the bridging computer and follow these steps:

1. **Select Start⇨Settings⇨Control Panel, and then select Network⇨ Internet Connections⇨Network Connections.**

   To create a network bridge, you must select at least two network connections that are not being used by Internet Connection Sharing (ICS) or the Internet Connection Firewall (ICF). Then again, if the computer is triple-homed and has three network cards, you can use one of the adapters to connect to the Internet while using the other two for bridging. Remember to make the bridge before enabling ICS; otherwise, it won't work.

2. **In the Network Connections folder, you will see a connection for each network adapter displayed. Click the connection for the wireless adapter.**

   In the Details pane, you should see that it's successfully connecting to your network.

3. **Click both network connections (when you have several segments, hold down the Ctrl key and click all the connections that correspond to all the LAN segments you want to bridge), right-click one, and select Bridge Connections from the menu.**

   You should see the message `Please wait while Windows bridges the connections....`

   After the bridge configuration is complete, Windows creates a new connection called Network Bridge (Network Bridge)#, where # is a Windows-generated number.

4. **Right-click the Wireless Network Connection icon in the Network Bridge section of the Network Connections window and select the Properties option to display the Network Bridge Properties dialog.**

5. **Click the Wireless Networks tab. Configure your wireless connection to match the access point you will bridge using Preferred Networks from the dialog. You should move it to the top of the preferred networks list.**

6. **Click OK.**

Troubleshooting a wireless bridge

First, the bridge will not work unless the computer acting as a bridge is turned on and working.

Second, from the Network Connections window, click both network connections and right-click one, and select click Repair from the menu. This may fix your bridge.

Where it looks like you made the bridge, but it doesn't actually pass network traffic, you should ensure that your wireless adapter supports promiscuous mode. Windows XP may have reported that your adapter does, but it may not. Unfortunately, this is a big problem with many wireless cards on the market right now. You can manually force the adapter into compatibility mode. To set your wireless adapter to ForceCompatibilityMode, follow these steps:

1. **Open a command window (click Start⇨Run), type** cmd **or** command, **and then click OK.**

2. **In the command window, type** netsh bridge show a.

3. **Note the number assigned to the wireless adapter and type** netsh bridge set a *number* e, **in which you substitute the number displayed in the previous step for the word** *number.*

4. **To double-check that you set the wireless adapter correctly with Force Compatibility Mode enabled, type the** netsh bridge show a **command again.**

You can also contact the manufacturer or check the manufacturer's Web site for more information about possible updates for your card.

Using Wireless Switches

One of the biggest growth areas in wireless is wireless switches. Most of the big players provide products in this market segment. For example, Airespace (www.airespace.com), Aruba Wireless Networks (www.arubanetworks.com),

Chantry Networks (www.chantrynetworks.com), Cisco Systems (www.cisco.com), Extreme Networks (www.extremenetworks.com), Gateway Inc. (www.gateway.com), Legra Systems (www.legra.com), Symbol Technologies (www.symbol.com), and Trapeze Networks (www.trapezenetworks.com/homepage.html) all have switch products. In addition, you can find a wireless LAN switch blog at www.wlanswitch.com.

These intelligent devices will change how you configure and manage your 802.11 networks in the future. Management features of a switch include class of service, quality of service, switched Virtual LAN support, and robust security. Class of service allows you to centrally manage and control access rights and features. For instance, you can turn off Instant Messaging and other peer-to-peer applications in certain locations. Multiple VLAN support allows you to centrally configure different access rights to different groups on the same network, a feature that you won't find with current access points. You also have the capability to manage hundreds of distributed nodes from a central location. A centralized system also allows you to run numerous security alternatives, including WEP, Kerberos, 802.1X, TKIP, and RADIUS. You can find references for wireless switches at www.nwfusion.com/topics/wirelessswitches.html.

BROADbeam (www.broadbeam.com/index.asp) purports to have seamless switching between WWAN and Wi-Fi networks. IntelliSwitching supports 802.11a/b/g, GPRS, GSM, 1xRTT, and newer networks, such as EVDO and EDGE.

One last thing on infrastructure devices: You may have heard about LWAPP (lightweight wireless access point protocol). *LWAPP* is an open protocol that defines the exchange of control and data traffic between lightweight access points and WLAN appliances. The control component of the protocol provides such functions as secure firmware download, device configuration, client session management, and tunnel setup. Using an open protocol like LWAPP ensures that you can design, deploy, and redeploy wireless networks that are interoperable and non-proprietary. LWAPP may not survive because Cisco has a proprietary competing protocol. Wi-Fi Planet has a good tutorial on LWAPP at www.wi-fiplanet.com/tutorials/article.php/3067751.

Chapter 15

Dealing with Network Throughput Issues

*T*o ensure that your wireless network runs effectively and efficiently and remains fast enough to keep your user community happy, you need to perform maintenance. A network sometimes hiccups if you are not paying attention to it. For many of you, performance issues should be relatively minor because you do not have a large number of wireless users and are not pushing tremendous bandwidth. Others, though, may be stretching the limits of their wireless budgets and will need to ensure that the performance remains suitable to their needs.

Watching Traffic

You have seen those people sitting at a roadside with a clipboard in hand, haven't you? They are watching traffic and counting cars. You sometimes need to do the same thing on your network, only instead of counting cars you count the number of packets crossing the network. You also watch traffic to learn how your network is handling the load provided by all your users at any given time. Through traffic watching, you can determine network performance, see data as it crosses the network, and perform network analyses.

Estimating network performance

A lot of things can negatively impact network performance, from poor device drivers to competing traffic to inconsiderate users downloading gigabytes' worth of MP3s on your network. All this makes for poor relations between the users and your technical staff. You need a method of determining that traffic and balancing sufficient load with your business needs.

To estimate the performance of your network, you need to understand the traffic that it will sustain. Are your users able to connect to the Internet and download MP3- or AVI-type files? Are network people using the wireless spectrum to download large patches and configuration files? How many users are on the network at a given time and what are their main job functions?

The performance of your wireless network depends on factors such as distance to an access point, structural interference of buildings and walls, and placement and orientation of devices, especially antennae. You really need expert advice to do this well. Sites such as `www.csm.ornl.gov/~dunigan/netperf/netlinks.html` can provide you with tons of detailed information on performance issues and calculations. Another interesting site is the Cooperative Association for Internet Data Analysis (`www.caida.org`), which offers specialized advice on Internet network traffic analysis. You might use this to determine the speed of your Internet connections.

You can use a rough formula, though, to calculate an estimate of traffic load on your network. Appendix C contains a table that provides frequencies and their data rates. Using 802.11b as an example, you see that data transfer can occur at up to 11 Mbps. Of course, the likelihood of you achieving anywhere near that speed is remote, so taking a conservative estimate of 5 Mbps, you can begin to calculate traffic load. Next, you need to know what you might be using over the network, such as e-mail or file transfer. If you are transferring a 1MB file, then divide that by 5 Mbps to get a transfer time of about 200 milliseconds (ms), assuming nothing else is going on. E-mail or other traffic may only consume perhaps 100 Kb, or roughly 10 e-mails for each megabyte. So, all things being equal, you can do a very rough estimate by deciding how many e-mails and file transfers will occur on the wireless network and then adding the number of users who might be connected to determine a threshold. You can use similar numbers for your 802.11g or 802.11a networks. But this is so elementary that it might not give you any real basis for determining overall performance.

To really get anywhere using real statistics, you need some form of toolkit. You can purchase network simulation tools for this task, such as OPNET Modeler (`www.opnet.com/products/modeler/home.html`) or their ServiceProvider

Guru. If these are too pricey, perhaps Dummynet (`http://info.iet.unipi.it/~luigi/ip_dummynet`), a free BSD-based product, might be useful. A good thing about this software is that you don't need to install BSD to run it; it comes on a bootable floppy disk. Plug it in and begin testing your bandwidth. Okay, it isn't quite that simple — you may need to add your wireless network adapters.

Other tools include the AirMagnet Handheld by Airmagnet, Inc. (`www.airmagnet.com`), which runs on Pocket PC devices. This tool can detect and send out alerts for over 80 wireless security and performance conditions. It also offers built-in tools for site surveying, connection troubleshooting, and coverage mapping. All that and you can wander around with it in your back pocket. Naturally, they also offer a version that runs on a laptop, for those of you with other needs or without Pocket PCs.

Another tool, Fluke Network's OptiView Series II Network Analyzer (`www.flukenetworks.com/us/LAN/Handheld+Testers/OptiView/Overview.htm`) not only analyzes the traffic, but also offers traffic generation capabilities, so you can flood the network and see how it responds.

If these do not appeal to you, try Airopeek (`www.wildpackets.com/products/airopeek`), which does a similar level of performance analysis as the others, analyzing signal strength and channel and data rates. You see in Chapter 16 how to use Airopeek to discover rogue APs. Windows NT Magazine (`www.winnetmag.com/Files/25953/25953.pdf`) offers a long list of such analyzers along with some general information about them. They include more of the high-end versions than we do in this book; so if you are flush with cash and think you need something stronger and more powerful, check it out.

With these tools, you want to find out how busy your network is at any given point. You do this by checking the traffic throughout a given time period and determining whether it meets your expectations. What expectations, you say? Well, that depends on you and what the wireless network is used for in your business. Is it a mission-critical application network? Is it merely offering a few Tablet PC users access during boardroom meetings? Do customers rely on it? All these need consideration to determine whether you care if the network gets busy and bogs down. Hopefully, you answered these questions when you developed your plan. You did develop a plan, didn't you? (If not, hurry to Chapter 2.)

To determine whether your network is operating at sufficient capacity, you can use CommView for WiFi from Tamosoft (`www.tamos.com/products/commwifi`), which is a wireless network packet analyzer. This tool is specific to wireless networks and offers many capabilities besides packet sniffing. One of its features is statistical analysis, which you can use to determine how busy your network is at any given time. Running this over several different time periods in a week can provide you with valuable information. You must know where you are in order to know where you are going.

When CommView for WiFi is running on your machine, it places the adapter in a passive mode. This means it cannot connect to the wireless network as a functioning client, so you cannot perform your regular business while also running the program. This is unfortunate, but setting up a machine specifi-cally for monitoring is not necessarily a bad thing. The installation is fairly straightforward, like most Windows software these days. Once installed, it offers a number of options, as you can see in Figure 15-1.

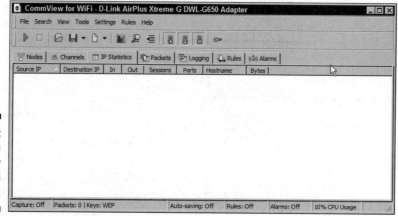

Figure 15-1:
Viewing the
CommView
for WiFi
main menu.

We discuss many of the settings later on in this chapter. For now, if you select View⊃Statistics, you see a page like that shown in Figure 15-2. This is where you can determine how well your network is running. It offers a number of options.

As you see, the Statistics menu offers Packets per Second analysis as well as Bytes per Second. The Bytes per Second can be changed to show Bits per Second. For each of these fields, the program shows the current average. Using this, you quickly see the overall impact your users are having on the network and can determine whether that impact is high or reasonable.

Within the Statistics page, there are seven tabs to select from, starting with the General tab that appears when you first open the statistics page. This tab offers the overall statistics, as mentioned previously. The next six tabs are shown in Table 15-1.

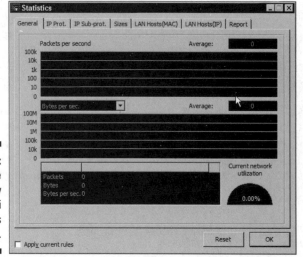

Figure 15-2:
Viewing the
CommView
for WiFi
Statistics
menu.

Table 15-1	Options Available in the Statistics View
Tab	_Description_
IP Prot.	This tab shows you the IP protocols.
IP Sub-prot.	In this tab, you see the other protocols, such as FTP and HTTP.
Sizes	Here you can easily see the packet sizes in use across the network.
LAN Hosts (MAC)	This shows the hosts on your system using their MAC addresses.
LAN Hosts (IP)	This shows the hosts on your system using their IP addresses.
Report	On this tab, you can set the parameters for your reports.

All these can be used to provide a fairly detailed view of your network, show-ing you trouble spots and overall utilization.

You cannot obtain data if the system is using WEP or WPA unless you add the proper keys because all packets are being encrypted. You add the keys to CommView for WiFi by selecting Settings➪WEP/WPA Keys and entering the keys in the space provided.

To start using all these tabs, you need to begin capturing packets so you can obtain some actual data. After you identify and input the proper keys, you need to start the capture process. Simply follow these steps:

1. **Open the CommView program if it is not already open.**

2. **Click the Start icon, or select File⇨Start.**

   A new screen called Scanner appears. This screen locates the wireless networks in the vicinity. In the Scanner section, click Start Scanning.

3. **The program will scan all channels for wireless signals and show them to you under the Access Points and Hosts section. Selecting one of the networks shown produces details about that network under Details. You see this in Figure 15-3.**

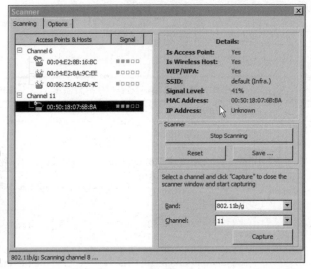

Figure 15-3: Viewing the CommView for WiFi Scanner page.

4. **Choose one of the networks and click Capture.**

   CommView begins to capture packets.

5. **Select View⇨Statistics to see how your network is handling the bandwidth load.**

   Another window shows the current data from the network you chose in Step 4. We chose a very large download from Microsoft, and in this example, we are using only one machine on the wireless network. You can see from Figure 15-4 that this creates a bandwidth load of about 4 or 5 percent (the figure is showing 4.6 percent). With a few more users on the network, each downloading files or sending e-mails, this small network will quickly be overloaded.

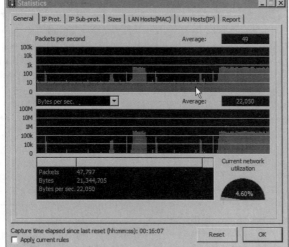

Figure 15-4:
CommView
Statistics
page
showing
utilization
figures.

As you can see, the tool allows for some useful data collection. In the other tabs, you can parse this data in a number of ways. In the IP Prot. tab, you see the number and type of packets (TCP, UDP, etc). In the IP Sub-prot. tab, the data is divided by the lower or sub protocols in use, such as HTTP, FTP, or POP. This can be especially useful to help you determine what your users are doing with the bandwidth. You can review the use of the other settings using Table 15-1.

6. **You can run a report using the Report tab and provide details in either HTML format or comma-delimited format depending on your needs**

   This enables you to produce an informative report for your management on overall performance of the network. Stop the program at any time by selecting File⇨Stop Capture.

The program offers a solid method for determining overall network performance at any given time. Running it at different times of the day and different days of the week and capturing the results in logs enables you to compare the data over time periods that you might feel are busy or indicative of the overall state of your network. Now you can determine whether one particular user is abusing the bandwidth, or whether a particular protocol is being heavily used, and take appropriate action.

You might also use the data gathered in this program to ensure that staff are abiding by any policies and standards you might be enforcing across your network. Chapter 10 discusses the types of standards you may want to use.

Sniffing your traffic

It's not polite to sniff in public, is it? It may not be polite to sniff your network traffic, either, but there are sometimes good reasons for doing that. You can look into packets and see what is happening. You can check for cleartext passwords and use that information to press for changes to systems still using such weak authentication. Other reasons include checking for wrong syntax of http requests or POP3 and ftp commands, or seeing what ports an application is using.

We use packet sniffers with clients on a regular basis when they need to allow an application to pass through a firewall but don't know which ports are needed. Sniffing the packets while the application runs is a simple way to determine that. We can recall one instance in which a service provider was confident that a particular application only needed one specific port to be open on the firewall, and was therefore not at risk. Using a packet sniffer, we discovered that the application actually opened different ports each time it ran, meaning we would have to open the entire range between our client and the other organization. This was just not acceptable, and we proved it with the sniffer. A newer version that acted properly eventually resolved the issue, allowing us to permit one open port and no more.

There are other reasons for using such applications. We discuss a few of them in previous chapters in discussions about hacking. We also provide you with a number of such tools in Chapter 17.

So how do you use a network sniffer? Continuing on with our example of CommView for WiFi, you select some of the other tabs shown on the main page. The following steps show you how to view data and other information found in a network packet.

1. **Click the Start icon or select File⇨Start. A Scanner screen appears. This screen locates the wireless networks in the vicinity. Under the Scanner section, click Start Scanning.**

   The program scans all channels for wireless signals and shows them to you under the Access Points and Hosts section. Selecting one of the networks shown produces details about that network under Details. (Refer to Figure 15-3.)

2. **Choose the network you wish to view, if more than one choice is available, and select Capture.**

   The program starts capturing packets.

3. **Click the Packets tab.**

   A screen will appear looking something like the one in Figure 15-5. Note that by dragging the mouse over the lines separating each section of the page, you can resize each section.

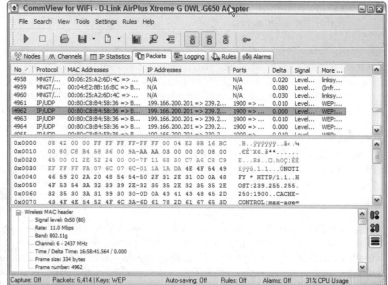

Figure 15-5: CommView showing packet details.

Three sections are shown:

- In the first section, you see each packet on one line with high-level information about it, such as the protocol, MAC address, IP address, the ports in use, and other fields. This alone provides enough information for tracking rogue applications to determine what source and destination ports they require.

- The second section shows the actual data within the packet. It is here you will see cleartext passwords when any are passing across the network, as well as any other information, such as Web sites being visited or file transfers.

- The third section provides detailed information on the actual packet, delving deeply into each one to show the SSID, WEP parameters, the band (a, b, or g), the channel, and a whole pile of other information. This section is only for network administrators who truly understand how TCP/IP works and can make sense of things like the ACK and SYN and ARP response. If you dig around, you'll find the BSSID and other useful data you should recognize from the various chapters in this book.

4. **When you have collected a reasonable amount of information, stop the collection by selecting File⇨Stop Capture. You can then save this data to a file for later viewing and analysis using the options found under the File menu.**

Don't let your network packet capture run for hours on a large network without checking to see whether you need that amount of information and ensuring you have enough hard drive space to hold it. It will quickly amount to tens of megabytes. It may also considerably increase your CPU usage and make the application less responsive. Consider filtering out packets you don't need for your analysis.

You see from these steps that the amount of data collected and the detail you can get from each packet is prodigious. You may want to read the book, *TCP/IP For Dummies,* 5th Edition, by Candace Leiden and Marshall Wilensky (Wiley), to find out more about this protocol.

We warned you that the amount of data you can collect can be huge. You may want to filter out those packets that aren't useful to the purpose of your collection. If all you want is statistical information, the green histograms, pie charts, and hosts tables, then use the Suspend Packet Output menu command, which allows you to collect statistical data without real-time packet display. You do this by selecting File⇨Suspend Packet Output after selecting the Start Capture option. This stops showing the packets, but keeps the statistical information for your charts.

You may want to select the Rules tab and then select options that will limit what is collected. The options on the left side allow you to select an impressive level of detail. You can see from Figure 15-7 that you can select traffic going to or coming from only certain MAC or IP addresses. You can specify specific ports to collect only certain application data, like FTP (23) or HTTP (80). You could also capture packets containing certain text information. This could be very useful in an investigation following up complaints of sexual harassment or other inappropriate use of your e-mail system. Naturally, you need to be sure that you follow any laws governing such access, and that you do not cross any privacy boundaries.

In Figure 15-6, you see that we have selected only ports 21 and 23 because we want to know what Telnet and FTP sessions are crossing the network.

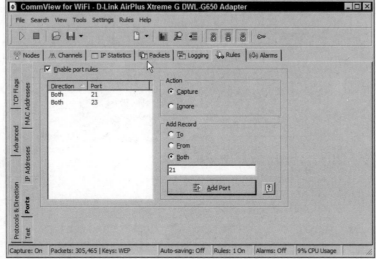

Figure 15-6:
Setting
CommView
to collect
specific
packet
information.

This merely touches on the use of this powerful tool, and we recommend that you study the documentation intensely to discover its full potential. Whether you use CommView or any of the other fine tools available, learning the details will allow you to respond quickly and effectively to any need you may have in your business.

One of the useful items that we will mention is the ability to reconstruct a session. This is useful because you certainly won't want to wade through every packet one by one, trying to see specific Web site or FTP session details. By right-clicking on the initial packet, you can select the Reconstruct This TCP Session option. You see this option in Figure 15-7.

If you select that option, the program reads all the packets pertaining to that session and provides you with a clearer look. You see the results in Figure 15-8. Note that you can modify the results to appear in ASCII (shown), HTML, or other display types depending upon your need. When you view FTP, Telnet, or Web site logins, or even that rogue application, this brings it all to bear and allows you to see the big picture.

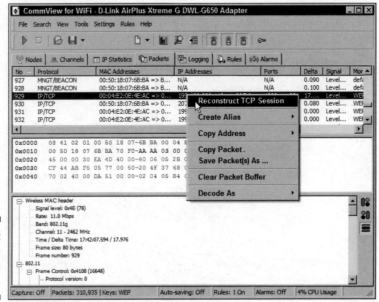

Figure 15-7:
Recon-
structing
a session.

Figure 15-8:
The
resulting
recon-
structed
session.

Notice that the Web page we visited is www.msn.com. You will see other information, of course. This is a powerful capability and is not to be underestimated.
These tools offer you the ability to manage and monitor your network effectively, and they belong in all companies' toolkits.

Traffic management and analysis

What do we mean by *traffic management?* Are we suggesting you enter an intersection and begin directing cars? No. We mean ensuring that your network functions well. The main goals of network management consist of the following:

- ✔ Improving network availability
- ✔ Centralizing control of the network components
- ✔ Reducing complexity
- ✔ Reducing the operational and maintenance costs

A network management system can reduce the cost and complexity of networks by providing integrated tools, allowing the network manager to quickly isolate and diagnose network issues before they become a major nuisance. Typically, it provides an ability to do this from a central location, removing network administrators' need to roam around in order to see and resolve issues.

The general areas network management systems deal with include those shown in Table 15-2.

Table 15-2	Key Network Management Functions
Function	*Description*
Fault management	This consists of detecting, isolating, and correcting any abnormal network operation. It includes getting the fault indication, determining the cause, isolating it, and performing corrective action.
Performance management	This consists of the tools used to recognize performance issues causing problems. It includes the ability to monitor the network for acceptable performance and collect and analyze statistics to help prevent future issues.
Configuration management	These include configuring and maintaining the network components.
Accounting management	This area involves measuring network utilization parameters to allow you to regulate each user's network use appropriately.
Security management	This encompasses all activities involved in controlling and monitoring access to the network.

Performing all this is a task your network people are charged with, and how they do it determines how well your network runs. You can use tools like Ipswitch WhatsUp Gold (www.ipswitch.com) or one we have used recently called SolarWinds Network Management Tools (www.solarwinds.net). You find a list of different vendor products in Chapter 16.

Using tools like Ipswitch WhatsUp Gold allows you to map out all the devices on your network and monitor them for availability, as well as monitoring individual services such as HTTP, DNS, or SMTP, or monitoring such things as disk space or memory utilization. Knowing that an object is having difficulty, however, requires notification, and the product performs this in many different ways. It can send a message to a pager, send an e-mail, or issue a pop-up on a console. Like CommView, this product can be used for performance statistics, reports on availability and errors, and a host of other options. Combining Ipswitch WhatsUp Gold with the efforts of a company called Wavelink Corporation (www.wavelink.com), you can use the product across both wired and wireless networks.

Organizations today rely heavily on such management tools to help ensure that their networks remain functional, and for quickly detecting and resolving problems. You should be using these tools on your network, as well.

Outsourcing your network management

If you outsource your network management, you need a service level agreement that indicates the precise degree of network availability and bandwidth utilization that is expected and over what time frame. A *service level agreement* (SLA) is a written agreement between your service provider and your company that clearly outlines the expected performance level of network services. This agreement should include specific metrics agreed upon by both parties. The values set for the metrics must be realistic, meaningful, and measurable. That data might include

✔ Interface statistics collected from the network devices, such as number of packets and ignored or dropped packets

✔ Size and type of network devices in use, including number of access points, stations, and switches

✔ Bandwidth utilization statistics

✔ Emergency response times and equipment upgrade or patch management implementation time frames

Using distinct, measurable, and quantifiable numbers increases the chance that you and your service level provider will be keenly aware of what is happening on the network and stick to the prescribed rates. Don't forget to include security metrics as well.

We have worked with an organization whose SLA was pitiable in its lack of distinct and measurable security metrics. This was to the point at which the firm was ripe for being taken advantage of, given that it would have no leg to stand on where opinions on measurement differed enough to impact the company in a negative way. For example, on a simple SLA that a service provider issued, they stood to be inundated with security audits because they placed no restrictions on their largest customer on how many audits they could request in a year. Typically, a service level agreement will spell out a reasonable approach, including using a standard audit that all customers would see, rather than specific ones for each customer.

Ensure that when your network is outsourced, your SLA is prepared with all your needs in mind and offers reasonable and qualitative metrics for measuring success.

Monitoring the Network for Trouble Spots

One key thing to look out for in your wireless network is rogue access points. You can do this by using a number of the management tools we mention. In CommView for WiFi, you use the Alarms icon on the main page. Other items you can look for include unknown IP or MAC addresses. These require more work, however, because most organizations use DHCP and not static addressing and few organizations know all the MAC addresses it uses. If you do know all of the MAC addresses your company uses, however, you can set alarms to go off when aberrations occur. Other uses include setting the alarm to look for bandwidth hogs and taking action when you find excessive use.

To scan for rogue access points, you need to know the MAC address for each access point on your network. Armed with this information, follow these steps:

1. **Open the program and select the Alarms tab. Then click Add.**

   You see a screen like that shown in Figure 15-9.

2. **Select the check box next to Rogue APs.**

3. **You need to configure the alarm. Click Configure.**

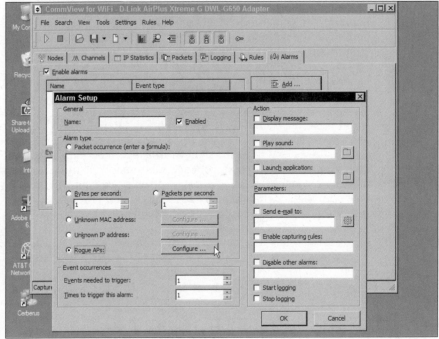

Figure 15-9:
Configuring
CommView
to detect
rogue
access
points.

4. **Enter the MAC addresses of your access points, and then click OK.**

5. **At the top of the page, enter a name for your rule in the Name field.**

6. **Select the type of action you would like to occur using the items listed on the right side of the page, and then click OK.**

   For instance, check the box for Display Message and enter a message such as **Rogue AP Detected**. After you click OK, you see your rule listed along with a check mark to indicate that it is active.

7. **If another access point is running on the channel you are scanning, your event is triggered and you see your message, as shown in Figure 15-10.**

This ability alone is a good reason to purchase CommView or other similar tools. While we set the event to trigger a message on the console running CommView, recall that you can send an e-mail, play a sound, or do any number of other things to attract attention.

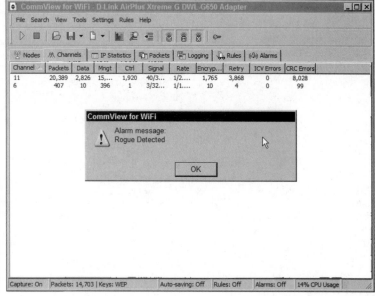

Figure 15-10:
Commview
detecting
a rogue
access
point.

Chapter 16

It's Ten O'Clock: Do You Know Where Your Access Points Are?

A big part of managing and protecting your network is knowing your network. Identifying your 802.11 and 802.15 gear will help you understand the magnitude of your problem. Many companies have emphatically stated that they had no wireless networks, only to find out they did. This chapter is for those who acknowledge that they have wireless networks installed (and for those who don't).

Discovering the Extent of Your Wireless Network

You have many ways to discover that you have wireless networks. You could send a survey out to your employees. We know that not everyone will respond to a survey. And those who do will probably not admit to having wireless if you have a policy against it. You could always participate in *management by walking around:* Take a stroll and look for access points and antennae. Look for people using computers in places that you know are not wired. Again, this is not 100 percent foolproof. If you have software inventory or configuration management software, you could look for client utilities. You could also supplement these methods with another automated one. After you have a wireless network up and running, you want to run a post-implementation site survey. To do so, you walk around with a laptop or handheld and do one or all of the following to discover wireless networks:

- ✔ Use the programs that came with your operating system.
- ✔ Use the utilities that came with your network adapter.
- ✔ Use war driving or network discovery tools.
- ✔ Use traffic management and analysis software.
- ✔ Use network management software.
- ✔ Use network vulnerability software.

Using programs that came with your operating system

As we point out in Chapter 6, you can use the built-in functionality of Windows XP and Mac OS X to discover networks. These operating systems are wireless network–aware. If you cannot remember how to use these utilities, go to that chapter and read up on using the tools to connect to a network.

Using utilities that came with your network adapter

Even though newer operating systems are wireless network–aware, your manufacturer will provide a utility to help you discover networks. In Figure 16-1, you can see the information you can gather by using the Client Manager that comes with ORiNOCO Silver and Gold cards. Use the pull-down arrows in the various boxes to change what you can display. The ORiNOCO tool also provides an excellent Link Test dialog box as well.

In Chapter 11, we show you another utility that comes with the Proxim 802.11a/b/g Gold PC Card. Try these manufacturers' utilities to test signal strength and more:

- ✔ **Site survey tools:** Discover networks, identify MAC addresses of access points, and quantify signal strength and SNR ratios.
- ✔ **Spectrum analyzer:** Find interference and overlapping channels.

✔ **Power and speed monitoring tools:** Monitor throughput and current connection capacity.

✔ **Profile configuration utilities:** Configure profiles for different networks.

✔ **Link status monitor with link testing functionality:** View packets, successful transmissions, connection speed, and link viability.

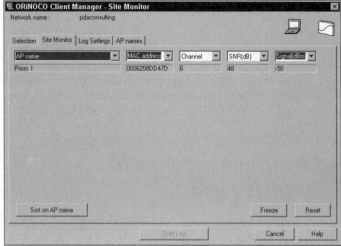

Figure 16-1:
ORiNOCO
Client
Manager –
Site
Monitor.

Use these tools but don't forget to use free network discovery tools, such as Boingo (www.boingo.com) as well.

Using war driving or network discovery tools

Several times in the book (for instance, Chapters 2, 5, 9, 15, and 17), we refer to *war driving* software. War driving software is the equivalent of the Swiss Army knife for network and security administrators alike. Of the many genre of this software, start with the following list:

✔ **AirMagnet:** www.airmagnet.com

✔ **AirSnort:** http://airsnort.shmoo.com

✔ **AirTouch Network Security System War Driving Kit:**
www.airtouchnetworks.com

- **dstumbler:** www.dachb0den.com/projects/dstumbler.html
- **kisMac:** www.binaervarianz.de/projekte/programmieren/kismac
- **kismet:** www.kismetwireless.net
- **MacStumbler:** www.macstumbler.com
- **MiniStumbler:** www.netstumbler.com
- **NetStumbler:** www.netstumbler.com
- **WaveStumbler:** www.cqure.net/tools.jsp?id=08

As you can see from the preceding list, several platforms and operating systems have software support. You have Mac OS X, Pocket PC, Windows NT/2000/XP, GNU/Linux, and FreeBSD support. NetStumbler for Windows and Kismet for GNU/Linux are the most popular of the network discovery genre. MacStumbler for Mac OS X and dstumbler for BSD are popular in their spheres.

Using traffic management and analysis tools

Wireless networks are broadcast networks, and broadcast networks are great for packet analyzers or sniffers. *Sniffer* is a trade name of Network General (which later became Network Associates). Unfortunately for Network Associates, the term *sniffer* became generic. Take out a tissue and cry for them. Okay, maybe not. But rather than offend anyone, we use the more generic name of *packet analyzer* (although some may prefer the term *protocol analyzer*).

Packet analyzers go beyond detecting the existence of a wireless network. By turning your wireless adapter into a promiscuous device, packet analyzers capture the frames you want, which may be all of them. You can set filters to determine what frames the packet analyzer will collect. With these tools, you can look at the contents of the captured frames right down to the bit level. You can fix existing problems or understand potential ones through analysis.

The following products support wireless packet analysis:

- **Ethereal:** www.ethereal.com
- **Berkeley Varitronics Grasshopper and Yellowjacket:** www.bvsystems.com
- **Epiphan CENiffer:** www.pocketpccity.com/software/pocketpc/CENiffer-2001-11-16-ce-pocketpc2002.html
- **Fluke Networks WaveRunner:** www.fluke.com
- **Gulpit:** www.crak.com/gulpit.htm

- **Netintact PacketLogic:** www.netintact.com

- **Network Associates Sniffer Wireless:** www.sniffer.com

- **Network Instruments Network Observer:**
  www.networkinstruments.com

- **Tamosoft CommView:** www.tamos.com/products/commview

- **WildPackets AiroPeek NX:** www.wildpackets.com

You can find more information about wireless sniffers at www.personaltelco.net/index.cgi/WirelessSniffers and www.blacksheepnetworks.com/security/resources/wireless-sniffers.html.

If you use the GNU/Linux version of Ethereal, it supports wireless packet analysis. The same is not true for the Windows version. With the Windows version, you have to grab the packets for analysis from the wired segment.

You can find further information on Ethereal in Chapter 17 and on AiroPeek NX later in this chapter. Berkeley's Grasshopper is a handheld, wireless, 2.4 GHz receiver that measures and displays RF power and narrowband receive signal strength indicator (RSSI) total channel power. What's more, you can use Grasshopper to measure packet error rate and channel usage. A dedicated tool such as Grasshopper is nice, but it may cost you as much as or more than a good laptop with packet analyzer software installed.

Several companies offer an appliance, as well. For example, Network Chemistry offers a Neutrino Sensor bundled with the freeware Packetyzer (www.networkchemistry.com/products/packetyzer) software, which allows you to capture and analyze 802.11 packets. The sensor captures all 802.11 packets and then forwards the captured packets over the wired network to Packetyzer for analysis and display. Packetyzer can decode WEP, WPA, LEAP, 802.1X, IPSec, and many other authentication protocols.

Using network management tools

You may already have some software in your organization that you can also use to locate authorized or unauthorized APs. Network management software can help you map out known and unknown devices.

Following is a short list of network management products:

- **3Com Network Director:** www.3com.com

- **AdRem Software NetCrunch:** www.adremsoft.com

- **Castle Rock Computing SNMPc:** www.castlerock.com

- ✔ **Cisco CiscoWorks:** www.cisco.com
- ✔ **Computer Associates UniCenter Application Performance Monitor:** www.ca.com
- ✔ **Enterasys Networks NetSight Atlas:** www.enterasys.com
- ✔ **HP OpenView:** www.hp.com
- ✔ **IBM Tivoli:** www.ibm.com
- ✔ **Ipswitch WhatsUp Gold:** www.ipswitch.com
- ✔ **Netintact PacketLogic:** www.netintact.com
- ✔ **Netplex Technologies SNIPS:** www.netplex-tech.com/software/snips
- ✔ **Opalis Software OpalisRobot:** www.opalis.com
- ✔ **SolarWinds Network Management Tools:** www.solarwinds.net
- ✔ **Symbol Technologies AirBeam:** www.symbol.com

The term *network management system* (NMS) means different things to different people. But basically, a good NMS provides what the ISO defines as FCAPS, which stands for the following:

- ✔ **Fault management:** Detection, isolation, and correction of abnormal network operation
- ✔ **Configuration management:** Configuration, documentation, maintenance, and updating of network components
- ✔ **Accounting or administration management:** Detecting inefficient network use, or abusing network privileges or usage patterns
- ✔ **Performance management:** Monitoring and maintenance of acceptable network performance, and collection and analysis of statistics critical to network performance
- ✔ **Security management:** Controlling and monitoring the access to network and associated network management information

Regardless of your organization's size, you will find that a network management system or tool effectively reduces the cost and complexity of your network. The NMS provides an integrated set of tools that allows you to quickly isolate and diagnose network issues. Sure, you can spend hundreds of thousands of dollars on IBM's Tivoli, but you will only spend hundreds of dollars to get WhatsUp Gold.

Using vulnerability testing software

If you have Internet access, chances are that you've used vulnerability scanners before. eEye (`www.eeye.com/html`), GFI (`www.gfi.com/languard`), Harris (`www.stat.harris.com`), ISS (`www.iss.net`), and Symantec (`www.symantec.com`) market some of the more popular commercial products. You may have heard of some of the more famous freeware ones: nessus and nmap. There are even database and application scanners, and now wireless scanners. Vulnerability scanners work by doing a point-in-time review, looking for known problems and reporting them to you.

Among the wireless vulnerability scanners are

- **AirMagnet:** `www.airmagnet.com`
- **ISS Wireless Scanner:** `www.iss.net`
- **WaveSecurity Wavescanner:** `www.wavesecurity.com`

Figure 16-2 shows you access points that Wireless Scanner found.

Click the Vulnerabilities tab, and you see a list like the one in Figure 16-3.

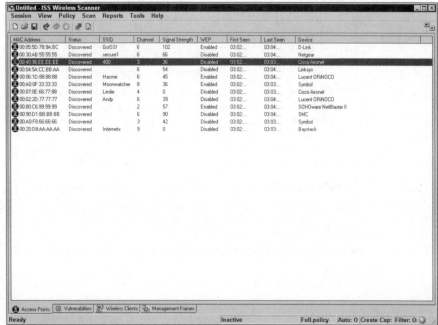

Figure 16-2: ISS Wireless Scanner.

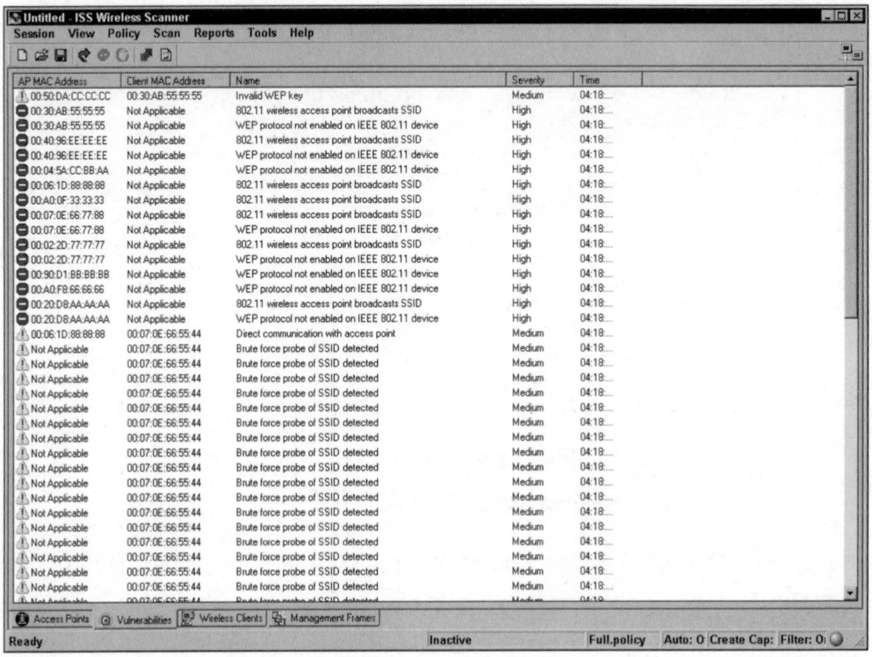

Figure 16-3:
ISS
vulnerabil-
ities tab.

You see two different icons in the view. The Yield (yellow) icon is a medium risk, and the Do Not Enter (red) icon is high risk. Don't know what the high risk vulnerability means? Simple. Just right-click any one, select What's this vuln? from the menu, and you see a description like the one shown in Figure 16-4.

The tools listed in all these categories cannot find any WAP or bridge that is not left turned on, so you need to supplement the tools above with one additional step. You need to check out the Web sites that document wireless LANs that other people have found. You can find these lists at

- **Nakedwireless.ca:** www.nakedwireless.ca
- **NetStumbler:** www.netstumbler.com/nation.php
- **Wi-Fi Zone:** www.wifizone.org
- **Wifinder.com:** www.wifinder.com
- **WiGLE:** www.wigle.net

We hope you won't find any of your access points on these sites!

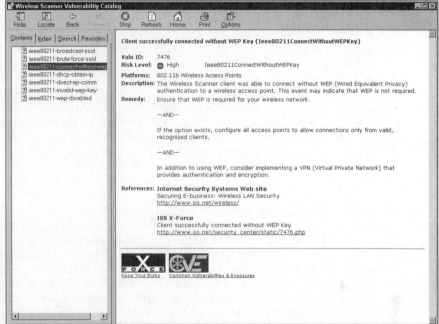

Figure 16-4:
Vulnerability
description.

Detecting Wireless Intrusion

Attackers can potentially intercept and decrypt sensitive data. (See Chapter 11.) *Crackers* also can gather sensitive data by introducing a rogue WAP into your coverage area. As well, your wireless networks are subject to denial of service (DoS) attacks, rendering the access point unusable to your clients. The attacks will only increase as the technology becomes more widespread. Without an intrusion detection or prevention system (IDS/IPS), you may have a difficult time determining when your WLAN is under siege.

You need an intrusion detection or prevention system regardless of whether you have deployed a WLAN. Intuitively, people believe that your wired network is at risk only when you deploy a wireless network. Unfortunately, any organization with a wired network also needs to monitor for WLAN traffic to make sure that the air surrounding them is not used against them. You need to concern yourself with your internal environment and rogue access points. Many times, organizations say that they don't use WLAN technology when the fact is that a number of employees, unbeknownst to management, have deployed WLANs.

A typical IDS/IPS will help you identify system and network intrusions and misuse by gathering and analyzing data. Traditionally, IDS were used for wired systems (host- or target-based IDS) and networks (network-based IDS). More recently, vendors have developed a wireless intrusion detection system (WIDS) for wireless networks. A WIDS can monitor and analyze user and system activities, recognize patterns of known attacks, identify abnormal network activity, and detect policy violations for WLANs. Wireless IDS gather all local wireless transmissions and generate alerts based either on predefined signatures or on statistical anomalies in the traffic. A wireless IDS is similar to a wired NIDS but has some features specific to WLANs. (*NIDS* is a *network-based intrusion detection system* where monitors gather and analyze all packets in the network and the system adjusts the security policy to block packets or improve efficiency.)

You can have centralized or decentralized wireless IDS. A *centralized* wireless IDS is usually a combination of individual sensors that collect and forward all 802.11 data to a central management system, where you can store and process the wireless IDS data. *Decentralized* wireless intrusion detection usually includes one or more devices called *sensors* that perform both the data-gathering and -processing and reporting functions of the IDS. If you have only one or two access points, you should select the decentralized method based on lower cost and less management. Installing many sensors with data processing capability is potentially cost-prohibitive for a small environment. On the other hand, you may find that management of multiple sensors takes more time in a decentralized model.

A WIDS can help your organization enforce its policy. Suppose that your policy states that clients must encrypt all wireless communications. With a wireless IDS enforcing, this is easy because you can continuously monitor traffic and ascertain whether any device is communicating without the use of encryption. You can use the WIDS to detect unknown or rogue access points. A wireless IDS can also detect the use of NetStumbler, Kismet, or similar software, helping to improve awareness of the threats to the WLAN. As well, a wireless IDS can detect some denial of service attacks, such as flooding authentication requests or disassociation/de-authentication request frames. A wireless IDS attacks another problem head-on by detecting the presence of MAC address spoofing. A wireless IDS also has the ability to recognize ad hoc networks. Finally, you can baseline traffic to identify when an intruder exceeds normal activity. You can then set thresholds and send an alarm to someone whenever any of the activities just listed occurs. In addition, a WIDS can monitor organizational compliance with policy by looking for the following:

- Unauthenticated traffic
- Default or improper SSIDs
- Access points, bridges, and stations operating on unauthorized channels
- Off-hours traffic

- ✔ Unauthorized data rates
- ✔ Unsupported protocols
- ✔ Unauthorized vendor hardware

Wireless intrusion detection systems work by continuously scanning an organization's airspace for evidence of an attack underway against the network.

Your IDS strategy should have two components: You should look for wireless-based attacks, and you should look for IP-based attacks. The wireless IDS should focus primarily on wireless attacks and not perform IP-based intrusion detection. To focus on IP-based attacks, you simply put a NIDS at the wireless AP choke point. Figure 16-5 shows how a WIDS and NIDS can work together.

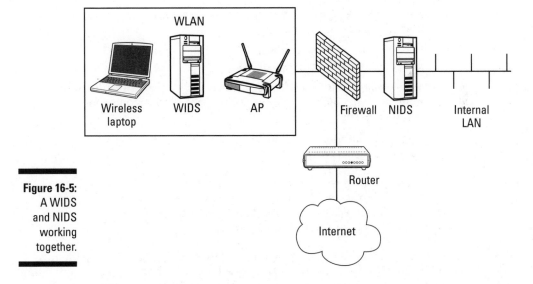

Figure 16-5:
A WIDS
and NIDS
working
together.

A wireless network requires both IDS technologies to provide proper visibility and coverage. The wired NIDS cannot detect any wireless-based attacks or wireless threats. Basically, a wired NIDS is useless against wireless attacks but can detect wireless-based attacks after they hit the wire. However, a wired IDS fails as an effective approach to detecting rogue wireless LANs because it cannot adequately identify

- ✔ Access points attached to the wired network itself
- ✔ Software-based access points
- ✔ Accidental associations
- ✔ Ad hoc networks

The NIDS can, however, detect IP-based attacks that the wireless IDS cannot. You need to aggregate events from both your wired and wireless IDS and correlate them to ensure maximum analysis.

There's a growing demand for wireless IDS/IPS, and there are many ways to meet that demand. If you're deploying a secure WLAN today and looking at your options for wireless IDS/IPS, you should consider the following:

- ✔ Your overall business needs for network security, not just for the WLAN.

- ✔ Your security budget and your business risk.

- ✔ Your level of in-house security and WLAN expertise.

- ✔ Your willingness to outsource security-related tasks. There are several Managed Security Service Providers (MSSP) doing IDS/IPS work.

- ✔ Your objectives for automated intrusion response and your need for related professional services.

- ✔ The type and span of wireless behavior that you want monitored (that is, security or performance or both).

- ✔ Your ability to find an MSSP that will work with you to understand and refine your needs, policies, escalation procedures, and incident response plans.

- ✔ History, reputation, and financial status of any company to which you may outsourced WIDS.

You can purchase a wireless IDS through a vendor, or you can outsource the monitoring, or you can develop one in-house. Only a handful of vendors currently offer a wireless IDS solution. However, you can program a wireless packet analyzer to look for potential intrusions.

For example, you can easily program a wireless packet analyzer to identify and locate rogue access points. You look for access points that are not yours, and then set a trigger for the event. After the analyzer trips the trigger, the program can send you an e-mail or run a program. For example, you can use AiroPeek NX to find rogue access points by following these steps:

1. **Choose File⇨New and then click the 802.11 tab.**

2. **Select Scan and click the Edit Scanning Options button.**

   Figure 16-6 shows you the Capture Options tab.

3. **Enable Channels 1–14.**

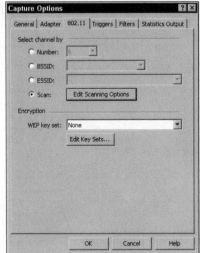

Figure 16-6:
Capture
Options tab.

You use Channels 1–11 only in North America, but that doesn't mean that a rogue access point is not using the other channels. AiroPeek doesn't transmit anything, so you won't violate any laws by listening on all channels.

4. **Click inside the Duration box.**

   You see a drop-down box.

5. **Select 100. Do this for all channels to do your channel scans at 100 ms. This is the value that WildPackets recommends for scans. Click OK twice.**

6. **Start the capture by pressing Ctrl+Y. Then select the Channels tab.**

   You see an arrow like the one in Figure 16-7. On your screen, the arrow will move up and down.

7. **Click the Nodes tab.**

   You see something like the view in Figure 16-8. The first line shows the ESSID (pdaconsulting), the second line shows the name of the access point (Linksys Group) and BSSID (8D:D4:7D), and the third line shows the individual workstations (D-Link Sys).

   You can assign a level of trust to any node. The three levels of trust are Trusted, Known, and Unknown. If you look in the fourth column, you will see that it reads Unknown all the way down. This is the default, but you can change it for an individual access point or workstation.

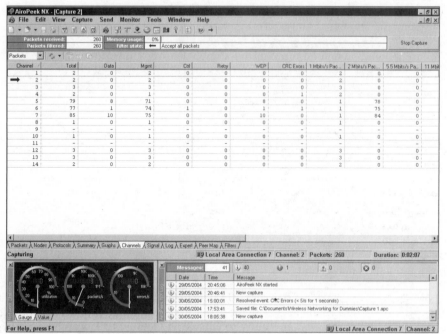

Figure 16-7: Channels tab of the capture.

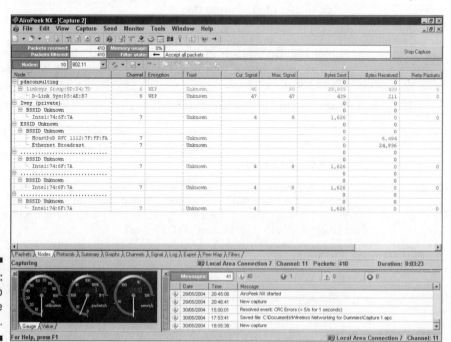

Figure 16-8: Nodes tab of the capture.

You use Unknown for the access points that you haven't previously identified or recognize. These are the potential rogues. Unknown is the default setting for an access point or workstation. Use Known for untrusted yet identified access points. These access points may reside in areas where you have no physical control. You use Trusted when you have identified the access point, and in fact trust it. You assign Trusted to the access points in your network after you perform an audit and determine that they were indeed trustworthy.

If you did a site survey, you should have a list of all the authorized access points and bridges. These obviously become your trusted nodes. If you didn't do a site survey, we suggest that you develop a list of trusted nodes before you start looking for rogue access points.

8. **Right-click any workstation or access point.**

   The menu shown in Figure 16-9 appears.

9. **Select Trusted.**

   The value in the Trust column for the one you just selected is now *Trusted*.

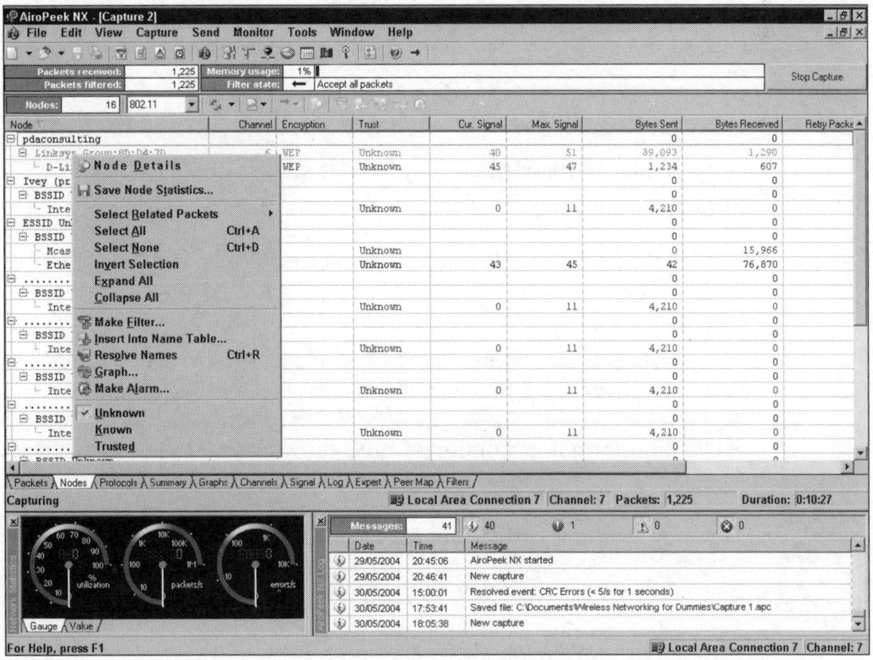

Figure 16-9:
Node menu.

10. **Click the Expert tab.**

In Figure 16-10, you see a caution symbol and the event Wireless Access Point – Rogue. AiroPeek treats any unknown devices as rogue. Your job is to follow up on all the unknown devices. You have the channel, but you will need a spectrum analyzer or a directional RF signal strength meter to help you find the physical device.

Figure 16-10:
Expert tab.

Layer	Event	Count
Physical	Wireless Access Point - Broadcasting S...	1
Physical	Wireless Access Point - Rogue	1
Physical	Wireless Channel Overlap	1,389
Physical	Wireless Client - Rogue	47
Physical	Too Many Physical Errors	2

If you're using Network Authentication, you can use your packet analyzer to look for failed authentications. These failures might mean an attempted access by an unauthorized person. You can come up with all sorts of similar tests when you start to look at the data that you are collecting.

You can find a list of WIDS at `www.pdaconsulting.com/idsprods.htm` or `www.securitywizardry.com/Wids.htm`. Commercial WIDS vendors include the following:

✔ **Airespace:** `www.airespace.com`

✔ **AirMagnet:** `www.airmagnet.com`

✔ **Aruba Networks:** `www.arubanetworks.com`

✔ **IBM:** `www.ibm.com`

✔ **Neutrino:** `www.networkchemistry.com/products`

✔ **Red-M:** `www.red-m.com`

WildPackets has an alpha of Omni Wireless Sensor, which it defines as a distributed spectrum analyzer. AirDefense also sells Bluewatch (`www.airdefense.net/products/bluewatch`). BlueWatch identifies all Bluetooth-enabled devices and their communications, allowing you to identify misconfigured devices or devices with no authentication or encryption.

If you're working with a more modest budget, try Snort-Wireless (`http://snort-wireless.org`) and WIDZ (`http://freshmeat.net/projects/widz/?topic_id+43,245,151,152`).

You might want to consider Arpwatch (`www.securityfocus.com/tools/142`), which monitors Ethernet activity and keeps a database of Ethernet-to-IP-address pairings. You use Arpwatch when you bridge your access points to the Ethernet segment. Arpwatch, running as a daemon, keeps track of the MAC address/IP address pairings as ARP replies pass through the network. It can alarm you when it notices anything unusual. If nothing else, you will have a nice log listing all the network clients.

Finally, what we see today is just the tip of the iceberg. Like the WLAN market itself, wireless IDS/IPS is relatively new. You can expect the market to change dramatically as it expands, matures, and inevitably consolidates.

A wireless IDS is only as effective as the individuals who analyze and respond to the data gathered by the system. A wireless IDS, like a standard IDS, can require vast human resources to analyze and respond to threat detection.

Incident Response and Handling

Your organization should develop a procedure to enable your clients to report suspected misuse of the wireless network, their accounts, or other misuse that they may have noticed. The trick is to develop the procedures before you have an incident; be prepared to respond before an incident occurs. Several companies like Coca-Cola and Tylenol have tried to do crisis management in real-time and found it just doesn't work. You must formalize the procedures to prevent ambiguity, and you must communicate them to everyone in your organization. Your procedures must include a method of notification. Whom do you call when you have a problem? The police? The press? The boss? In what order do you make calls?

The main goal of your incident handling process is to protect confidential, sensitive, or proprietary information. You also want to

- Assure integrity of (life) critical systems.
- Maintain and restore data.
- Maintain and restore service.
- Figure out how it happened.
- Avoid escalation and further incidents.
- Avoid negative publicity.
- Find out who did it.
- Punish the attackers.

You should consider a bunch of other things, too, such as a mandate and a response team. Section 5 of the Site Security Handbook (RFC 2196 at `ftp://ftp.rfc-editor.org/in-notes/rfc2196.txt`) provides an excellent starting point for developing your incident handling program.

In addition, you need to monitor security advisories. You can find generic ones like US-CERT (www.kb.cert.org/vuls), Sintelli (www.sintelli.com), and SANS @Risks (www.sans.org/newsletters), but you also want to keep apprised of vulnerabilities for your gear. For most vendors, this means that you have to monitor the generic vulnerability mailing lists or visit the site, frequently looking for updates. For example, you can find a list of security advisories for Cisco products and services at www.cisco.com/en/US/products/products_security_advisories_listing.html, and you can sign up for D-News from D-Link (www.d-link.com) by selecting D-News Sign Up.

Auditing Activities

Because WLANs use radio transmissions, they are inherently more difficult to secure than wired LANs. You must pair even the best passive defenses with an active defense for success. First, you must identify and stop attempted breaches. That is the point of the WIDS. Second, you must continuously monitor your airspace to ensure that people comply with your security policies. That is the point of an audit. Don't think that auditing is done only by auditors. An audit is a fundamental tool for establishing a baseline and understanding the nature of your network.

You should perform an audit of your wireless network before going live, after major changes, on a risk-based cycle, and when there are significant unplanned outages. At a minimum, you should plan an annual review of your network.

You can use automated tools to ensure compliance with your security policies. Tools are available for most platforms whether you favor Windows or Linspire (formerly known as Lindows). You can use a good wireless analyzer to monitor compliance with security policies, and also to identify, intercept, log, and analyze unauthorized attempts to access the network. Wireless packet analyzers can automatically respond to security threats in a variety of ways, making them ideal tools both for monitoring and for more focused analysis.

Real-time packet analyzers scan traffic on a network using preset triggers looking for security anomalies. These analyzers typically provide a set of security-related expert diagnoses, which may include

- Denial of service attacks
- Man-in-the-middle attacks
- Security policy noncompliance (perhaps an error of omission)
- Security misconfiguration (perhaps an error of commission)
- Rogue access point and unknown client detection

AiroPeek ships with a security audit template, which you can use as-is or extend or modify to meet your unique requirements. The template makes use of special filters, alarms, and preconfigured capture sessions to create a basic WLAN security monitoring system. The security audit template scans network traffic in the background, looking for indications of a security breach. When it finds one, it captures the packets that meet its criteria and sends a notification, keeping you informed of suspicious activity on your wireless LAN. To use AiroPeek security audit templates, you do the following:

1. **Choose File⇨New from Template⇨Choose.**

2. **Find the WildPackets directory on your system. Open the AiroPeek NX folder and the 1033 subfolder. Find the Security Audit Template folder. If you can't find it following these instructions, just search for the Security Audit Template folder.**

3. **Open the folder and open the Security Audit Template.**

4. **Choose File⇨New from Template, select Security Audit Template.ctf, and then click OK.**

5. **Press Ctrl+Y to start the trigger.**

6. **Click the Triggers tab. Select the triggers that you want to capture.**

In addition to the many tools discussed in this chapter, you should supplement your wireless audit arsenal. Test your system with the same tools that the crackers use. Of course, you will do it under controlled conditions. You can start with the following tools:

- **Cain & Abel (**www.oxid.it**):** Use this to "recover" passwords. Like L0pthcrack before it, Cain can sniff passwords off the network.

- **Cqure AP (**www.cqure.net/tools.jsp?id=09**):** Use this tool to set up a hijack.

- **ettercap (**http://ettercap.sourceforge.net**):** Use this tool to capture passwords and inject traffic. Most powerful tool of its kind.

- **HostAP (**http://hostap.epitest.fi**):** Use this tool to turn your workstation into an access point.

- **Lucent Registry Crack (**www.securityfocus.com/tools/2370**):** Use this utility to decrypt WEP encrypted key values stored in the Windows Registry.

- **NetScanTools Pro (**www.netscantools.com**):** Use this utility to perform ping sweeps and port scans.

- **prismdump (**http://developer.axis.com/download/tools**):** Use this to crack passwords by means of a wireless adapter with the Prism chipset.

- **Seque SoftAP** (`www.pctel.com/prodSegSam.html`): Use this to set up a hijack.

- **WEPcrack** (`http://wepcrack.sourceforge.net`): Use this to crack WEP encryption keys.

- **Win Sniffer** (`www.winsniffer.com`): Use this to grab passwords for FTP, HTTP, ICQ, IMAP, NNTP, POP3, SMTP, and Telnet logins.

- **WS_Ping ProPack** (`www.ipswitch.com`): Use this tool to perform ping sweeps and port scans.

When downloading audit tools, make sure that you can trust the source. Many Trojan horses have masqueraded as security tools. Remember what your mother told you about taking candy from strange people on the street.

Part V
The Part of Tens

The 5th Wave   By Rich Tennant

"Sure, at first it sounded great – an intuitive network adapter that helps people write memos by finishing their thoughts for them."

In this part . . .

This part is kind of fun and provides you with many tens of valuable nuggets of information for use in your pursuit of wireless networking. Much of this information provides additional depth to the previous parts and more places to go in your search for perfection. You find some great tools and ten awesome ways to protect your network (you cannot be too careful in today's networking world). Another resource is a list of ten great ideas on how to use your new wireless connectivity in business to obtain complete enjoyment and freedom now that you have wireless up and running.

Chapter 17

Ten Administrator Tools and What They Do

Knowing that you have a wireless network running doesn't resolve you of the need to administer it and keep it secure. In this section, we show you ten tools that can help you identify rogue networks, test the strength of your security, and find out what the world already knows about you.

Make sure that you check all software downloads with up-to-date antivirus software. Also, check with the vendor for latest releases or potential security exposures and its related patches. You may recall that the Blackice Firewall software recently suffered from a critical exposure. If it happens to respected organizations like Blackice, it can happen to anyone. For example, the recent Ethereal version 0.10.2 has multiple critical vulnerabilities.

Using Ethereal to Look at Traffic

If you're a command line weenie, you will just love the UNIX version of this tool. Originating in the UNIX world, Ethereal (www.ethereal.com) allows you to analyze the information traveling across your wireless network. Although it allows you to see information crossing your network, it doesn't actually detect wireless networks. You'll use some of the other tools in this list to do that. Of course, like so many tools these days, a Windows version allows those of us with a UNIX phobia to have the same capabilities.

Administrators use Ethereal when they experience network performance problems. Ethereal helps you pinpoint which network interface cards, cable segments, or protocols generate the most traffic. That information lets you determine whether you should upgrade the cabling, change a faulty network card, or track down a noisy application that is spewing out network packets.

If you prefer commercial products, you can obtain Commview (www.tamos.com). This tool performs the same functions as Ethereal but offers technical support and a few additional functions like Remote Agents. These Remote Agents allow capturing of data on remote network segments. This is useful if you are in, say, Toronto and want to monitor traffic in the New York office.

You also can use Ethereal to detect unencrypted passwords or other confidential data traversing your networks!

Stumbling on Networks with NetStumbler

After stumbling around in the dark for a while, you realize that something is missing. Light! Find out what is missing on your network. Looking for rogue wireless access points is a necessary evil today because it's so easy for those folks in accounting to add one without telling anyone. You know how those accountants just love to have fun!

NetStumbler, or one of its relatives, is one of the most useful and well-known tools available. Added to your arsenal, it identifies wireless networks while you wander around. Using a laptop or PDA, a wireless network card, and this program, you can roam around the neighborhood and find which of your neighbors is using a wireless access point. You can even find that rogue access point that those accounting weenies added. Barry has done this in Melbourne, Toronto, and various U.S. cities. Sweet. It is amazing how easy it is to detect wireless networks. It's 11 p.m. — do you know where your access points are?

Okay, so Barry detected a bunch of networks. Big deal. Well, it could be a big deal if they aren't secured. That means someone could use your network. A Toronto news item in November 2003 identified a man caught in his car literally with his pants down using a wireless network to download child pornography as he drove down a residential street. That traffic traveled across someone's network. Be sure it's not your network. Use the security options in your access point and know where your access points are at all times.

These tools do more than just detect networks, however. They also indicate whether security is enabled; the name of the access point vendor; and numerous details about signal strength, network noise, and MAC numbers. Naturally, you can save this data for later analysis. Figure 17-1 shows the Windows version of NetStumbler. Notice how it indicates the MAC address and SSID of the wireless networks that it's finding.

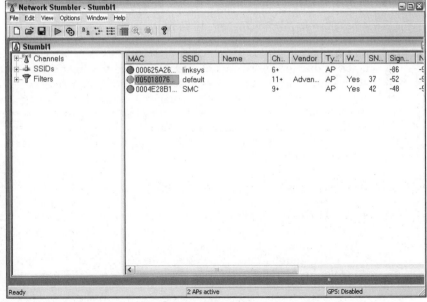

Figure 17-1:
The Net-Stumbler tool capturing wireless networks.

The various versions work the same yet allow for different platforms:

- ✔ NetStumbler (www.netstumbler.com) is the Windows version.

- ✔ MiniStumbler (www.netstumbler.com) works on PocketPC devices.

- ✔ MacStumbler (www.macstumbler.com) is for, well, Macintosh computers, of course.

✔ WaveStumbler (www.cqure.net/tools.jsp?id=08) is a console-based, 802.11 network mapper for Linux.

✔ Finally, iStumbler (www.istumbler.net) is another Mac OS application for finding 802.11b and 802.11g networks.

At least one of these tools belongs in your toolkit and should be used regularly. Stay one step ahead of the big bad world.

With Luck, You Can Find Networks with Kismet

The word *kismet* means *fate* or *fortune*. Although it would be nice to think the developer of Kismet arrived at that word because he felt he was changing fate with his work, alas — he merely liked the way it sounded.

Kismet (www.kismetwireless.net) is a wireless network detector and packet analyzer that primarily runs on UNIX systems. The big difference from other wireless tools in this list is that Kismet adds intrusion detection. This allows the user to detect potential intrusions in their wireless network. It can also detect other tools like NetStumbler and Wellenreiter (see the next section). Neat. Although it isn't exactly a replacement for commercial intrusion-detection tools like RealSecure or even AirDefense (www.airdefense.net), every little bit helps.

One distinct advantage, according to the developers, is the use of Rfmon rather than promiscuous mode. This is a special mode that reports all packets the wireless card sees, including management packets and packets from any network that the radio can see. Promiscuous mode software obtains data only from networks that it is associated with and doesn't see other networks' data. Thus, Rfmon arguably locates more networks than other tools. Get this tool if you're into UNIX and enjoy messing with non-GUI interfaces.

Of course, if you are a Mac OS X user, you need KisMAC (www.binaervarianz. de/projekte/programmieren/kismac), which is the version for Macintosh computers.

Surfing for Networks with Wellenreiter

German for *surfer,* Wellenreiter accurately describes the tool. We're goin' surfin', man! Hang ten! Find those wireless networks. You could use this tool surfing down the street on your skateboard, PDA in hand.

Wellenreiter (`www.remote-exploit.org`) searches out wireless networks and shows them to you, similar to NetStumbler and Kismet, but instead focuses on using handhelds like Zaurus and iPaq. This could be a handy tool to carry with you while you check e-mail or chat on ICQ with your Zaurus.

Using AirSnort to Obtain WEP Keys

Makes you feel like taking a deep breath of fresh air, doesn't it? Go ahead; no one's looking. AirSnort (`airsnort.shmoo.com`) runs under Linux, and requires that your wireless network card be capable of that Rfmon capability we talk about in the earlier Kismet section. This means network cards like Cisco Aironet and some ORiNOCO cards.

The big difference with this tool is that its main use is for breaking into secured networks. Don't use this tool except on your own authorized networks, or you'll run into trouble with the law in many states and provinces. It is a tool that recovers encryption keys. After you have the encryption keys, you can operate as if you were part of the trusted network and see all that lovely data.

AirSnort operates by passively monitoring transmissions and then computing the encryption key when enough packets have been gathered. It needs between 5 and 10 million encrypted packets in order to do that, but after it collects them, it takes only a few seconds to determine the WEP key. Yikes!

This may sound like a lot of packets, and for a really small business it might be. Look for that car parked outside for hours with the sneaky looking guy smoking cigarettes and tossing the butts out the window. For most businesses, however, collecting this many packets won't take long: a few hours on a busy network.

Why this can be done is a detailed affair and not for the faint-hearted. For the lion-hearted, it is described in the article, "Weaknesses in the Key Scheduling Algorithm of RC4," by Scott Fluhrer, Itsik Mantin, and Adi Shamir. You find it on the web using a search of the title. Numerous sites host it including `http://www.drizzle.com/~aboba/IEEE/rc4_ksaproc.pdf`.

Rooting Around with THC-RUT

THC is a short form for *The Hacker's Choice.* THC was founded in 1995 in Germany by a group of people involved in hacking, *phreaking* (tricking phone systems into providing free calls), and anarchy. Zounds! Sounds scary, doesn't it?

This group wrote the UNIX tool called THC-RUT (`packetstormsecurity.nl/ filedesc/thcrut-1.2.4g.tar.html`). This is another tool that administrators should add to their toolkits in order to manage their networks properly. Analyzing network traffic is cool, but it's only the tip of the iceberg. You also need to perform network discovery to know things like how many machines are connected and what operating systems they are using. THC-RUT ("aRe yoU There," pronounced *root*) is a toolkit that can do that. It gathers information by using a wide array of network discovery tools:

- ✔ Arp-lookup on an IP range
- ✔ ICMP-ping
- ✔ OS fingerprinting
- ✔ High-speed host discovery

The authors state that the tool is capable of discovering a Class B network within ten minutes. Use this tool to further your knowledge about the machines available on your wireless network.

Cracking Encryption with WEPCrack

WEPCrack (`wepcrack.sourceforge.net`) is a collection of Perl-based tools that allows you to attack the encryption algorithm used in wireless networks. It claims to be the first publicly available code that demonstrates the possibility of cracking WEP encryption, beating Airsnort by a few days.

It is composed of three main scripts that collect information, store it, and attack the results, obtaining the encryption key. The code is said to work with either 40-bit or 128-bit WEP.

Getting a MAC Address

Every business has an address, and network cards are no different. Each one has a specific and unique number: its *Media Access Control address.* CC Get MAC Address (`www.youngzsoft.net/cc-get-mac-address`) is a tool for finding these addresses. At the same time, it shows the computer's NetBIOS name. Network administrators can use it to find out who changed their MAC address, either because of a new network card or because of network spoofing attempts.

If you want to know your MAC address on a machine running Windows NT, 2000, 2003, or XP, just open a command prompt, type **ipconfig /all**, and look for the numbers appearing after the words `Physical Address`. If you have

Window 98 or Me, it's time to upgrade. Just kidding. You would use **winipcfg**. If you use UNIX/Linux, type **ifconfig -a** to get the same results.

Using this tool just automates something anyone can do, right? Well, yes and no. It duplicates the data gained from the earlier command, but it also allows you to run it across the network and see all MAC addresses and not just the one from your machine. If you collected all these addresses after ensuring that all your workstations and servers were powered on and running, you could then use that list as a comparison for later to determine whether any rogue workstations are on your network. This can help you identify intruders that may gain access over the wireless network. You also could use the addresses that you determine for MAC filtering. Most access points allow MAC filtering: You specify either those included or those excluded.

You may find that checking MAC addresses is a waste of time. A clever cracker can change his MAC address quite easily. If you don't want to play with your Registry settings in Windows, you can use Set MAC Address or SMAC (www. klcconsulting.net/smac) to change your hardware address to an acceptable one. Figure 17-2 shows the SMAC product. In the figure, Barry is changing the MAC address of his network interface card by keying in a new address in the field called *New Spoofed MAC Address*.

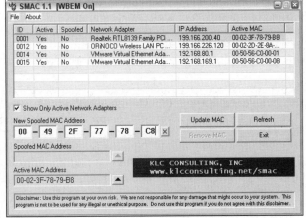

Figure 17-2: The SMAC tool with a spoofed MAC address.

Creating Sham Access Points with FakeAP

What if you faked having more than one wireless network? In fact, what if there are so many that finding the one you need is hard? This is the premise of FakeAP (www.blackalchemy.to/Projects/fakeap/fake-ap.html). This tool generates thousands of fake access points, allowing you to hide in plain sight. Although we are not usually advocates of security by obscurity, this tool does cause us to rethink that old axiom.

FakeAP is a Linux-based tool that uses Perl scripts, and its only goal is to try to obfuscate your real wireless network by placing as many as 53,000 fake ones around it.

We are not sure that it is the best approach to security, but it is an interesting concept. Time will tell whether it has any real value. Naturally, we advocate security using the formal methods available within your wireless access point, rather than relying on obscurity alone.

Let's Sneak a Peek with AiroPeek, Shall We?

No, we are not talking about being rude here. No peeking at exam time. AiroPeek NX (www.wildpackets.com) is a Windows-based wireless networking tool, or *sniffer,* that offers some enhanced capabilities. The tool provides the information-gathering aspects of NetStumbler, but it also offers features that ensure secure configuration of your wireless networks. This includes the ability to detect security items like rogue access, risky device configurations, Denial-of-Service attacks, Man-in-the-Middle attacks, and Intrusion Detection. Whew! An all-in-one solution. We're getting tired just thinking about it.

AiroPeek NX comes with a Security Audit Template that creates a capture window and then triggers a notification when any packet matches a specifically designed security filter. This allows the administrator to search for applications like Telnet or access points by using default — and therefore not secure — configurations. Unfortunately, the cost seems to be either a secret or an incredibly complex process; the AiroPeek Web site requires you to fill out a request form to get any idea of what it charges. Guess we can't take a peek at the cost, can we?

If AiroPeek doesn't suit your needs or your budget, go to www.personal telco.net/index.cgi/WirelessSniffers where you can find a comprehensive list of wireless sniffers.

Chapter 18

Top Ten Ways to Secure Your Network

*E*veryone has a top-ten list of wireless security vulnerabilities and controls. If you have not seen any, here is a sampling:

✔ www.npower.org/NPowerServices/NationalProjects/
   clip%20your%20strings.full%20handout.pdf

✔ http://arstechnica.com/paedia/w/wireless-security-howto/
   home-802.11b-1.html

✔ www.loud-fat-bloke.co.uk/w80211.html

Why should we be any different? This chapter focuses on ten significant vulnerabilities and compensating controls. You will find easy-to-follow advice for filling in some of the biggest holes in your wireless network.

Using the Highest Level of Encryption

We cover encryption extensively in Chapter 11. You get quite a lot of information about it. The first thing you discover is that encryption is optional. The original standard did not make encryption mandatory, so you must enable it. You can't use the highest level of encryption unless you turn it on, so turn it on!

The original WEP (Wired Equivalent Privacy) standard called for only a 40-bit key. As you recall, some vendors called this a 64-bit key. However, they use the first 24 bits for the initialization vector (IV). Each frame gets a different initialization vector; otherwise, when you and I send the same plaintext, we will get the same ciphertext. (The same cleartext, the same algorithm, and the same key would give us the same ciphertext.) Now, a 40-bit key is not nearly long enough. You need a key length of at least 80 bits. So, an enhancement to the standard allowed 104-bit (or 128-bit when you include the IV) key lengths. Generally, a 104-bit key is strong enough. But it didn't exactly work out that way. The problem lies in the implementation of the RC4 stream algorithm and not in the key lengths themselves.

WPA (Wi-Fi Protected Access) tried to correct some of the problems of the original WEP algorithm. For one thing, the initialization vector size was doubled. This meant we were less likely to get IV collisions, which help cryptanalysts break code. WPA also allowed longer key lengths, that is, 152 bits. This solved some of our problems, but we were still using the RC4 algorithm.

So, the next iteration introduced the Advanced Encryption Standard (AES). AES was based on the Rijndael algorithm. (See www.esat.kuleuven.ac.be/~rijmen/rijndael for more information on the Rijndael algorithm.) AES is a block cipher. Rijndael supports different key lengths (128, 192, and 256 bits), but IEEE 802.11i supports only a 128-bit key. It works on 128-bit chunks of data at a time (this is fixed as well in the standard), unlike RC4's stream cipher approach of encrypting one byte at a time (by XORing). AES has a variety of modes. The first implementation of AES used the Electronic Code Book (ECB) mode of AES. However, this mode has its problems as well. Suppose that you create text consisting of 64 A's. If the block size is 128 bits or 16 bytes, then using ECB produces identical ciphertext for the blocks of encrypted data. You may find some products based on the ECB mode, but you should avoid them.

In addition, there is an Offset Codebook Mode (OCB) of AES that provides message encryption and authentication in a single computation. OCB was the first mode selected by the IEEE 802.11i working group and they called it WRAP (Wireless Robust Authenticated Protocol). Just love these acronyms, don't you? At any rate, you should avoid WRAP or ECB or any other mode in favor of the CBC-MAC (Cipher Block Chaining-Message Authentication Code) mode or AES-CCMP (Advanced Encryption Standard-Counter Mode-CBC MAC Protocol), as it is now called. CCMP was adopted as the mandatory mode and the working group dropped OCB and ECB as mandatory, but you may find them in proprietary solutions.

So the choice is easy. Use the strongest encryption available to you. If all you have available to you is WEP with a 40-bit key, then use it. It is better than no encryption at all. Make sure you use the right mode and the longest key length. This is the easy part, figuring out what to do with the keys is the hard part!

Changing the Default SSID

We have told you several times about NetStumbler and its ilk. If you want to have some good clean fun, get in your car and drive around your neighborhood with an external antenna with good gain and NetStumbler running on your laptop. You won't need to travel very far (most likely you never have to leave your driveway) before finding an access point with the factory-set default SSID. In all probability, the access point owner will not have enabled encryption either. There tends to be a pattern here of not following the vendor's documentation.

Throughout this book, we try to dissuade you from thinking that an SSID constitutes a passcode of any kind. It is a network identifier, pure and simple. Because you already read that your access point broadcasts your SSID (unless you disable the broadcast), you realize that it's child's play to acquire your SSID. On top of all that, you will find sites such, as CIRT (www.cirt.net), that document the default SSIDs. So changing the SSID has little effect, but perhaps it is enough to convince someone that you are not an easy target. Pleasure crackers look for the low-hanging fruit. Dedicated attackers — well, that is a horse of another color.

Looking for Rogue WAPs

Generically, a *rogue* is a deceitful and unreliable scoundrel. A rogue wireless access point (WAP) is a deceptive mechanism used to trick a victim into believing — the deceit — that they are using one resource when they are actually using another — unreliable. Hence, rogue wireless access points are those that you or your organization have not authorized to attach to the wired network.

Network administrators, when using a wireless "sniffer" such as AiroPeek, sometimes get a rude awakening. They find many unauthorized WAPs. It is relatively easy to set up a rogue WAP. Anyone with about $100 can acquire the hardware for an access point. This is not a great barrier to entry. By installing a WAP on an established LAN, an individual can create a backdoor into your network, subverting all the hard-wired security solutions and leaving the network open to crackers and other, more benign, users.

You may notice a WAP hanging from the ceiling above your head, but it may take a while to find if it is placed in a discreet spot that allows maximum coverage. It may take even longer to gain unauthorized access to it if the networking staff takes care to turn off SSID broadcasts and disables SNMP or other management features, such as browser support. If the device has a built-in firewall that can block ARP broadcasts, it may last even longer.

But someone could set up a laptop running Red Hat Linux with HostAP (http://hostap.epitest.fi) and Airsnarf (http://airsnarf.shmoo.com) and leave it sitting around on a desktop where it wouldn't stick out. Even more insidious, Linksys has a WLAN adapter that you can plug into an existing RJ-45 Ethernet jack on any device. So someone with access to a hub, switch, or router can just plug it in and away they go. This Wireless-G Ethernet Bridge has a capacity of 43 Mbps using 802.11g. Whoops, there goes the neighborhood.

This dastardly individual sets his SSID to the SSID of your existing network and waits for people to associate. The cracker most likely will select a different channel than the one you are using; otherwise, your AP and the rogue will compete for clients. Clients associate with the AP with the specified SSID with the strongest signal. After a client associates, the cracker can monitor all communications through the rogue WAP. Of course, this person has his sniffer running and is grabbing usernames and passwords as fast as your employees can enter them.

In addition, this rogue may extend your existing wired or wireless network well beyond what you had originally intended. Again, you might as well pull some CAT5/6 cable into your parking lot and invite everyone to use it.

You need to get out Kismet or Wellenreiter and look for wireless access points. Better yet, use a wireless packet analyzer such as AiroPeek or Wireless Sniffer. Products like Enterasys, though pricey, have rogue access point detection built in. If you have the money and the expertise, use a spectrum analyzer to find sources of electromagnetic radiation (an antenna is an example of an intentional radiator).

But really, the first step to fighting rogue access points in your organization is to create the right environment. You should develop a policy and articulate it to everybody in your organization. The policy must specify what kind of use is acceptable regarding client stations and wireless. Obviously, connecting an unauthorized access point to a wired network is not acceptable, and you must spell that out in the policy. Make sure everyone understands the ramifications of noncompliance with any policy you develop.

Disabling Ad Hoc Mode

Ad hoc networks allow synchronization with network systems and application sharing among devices. Although most WLANs operate in the infrastructure mode and architecture, another topology is also possible. This second topology, the ad hoc network, allows mobile devices in the same area (really, in the same room) to easily interconnect. In this architecture, client stations group into a single geographic area and can inter-network without access to the wired LAN (infrastructure network). We refer to the interconnected devices in ad hoc mode as an independent basic service set (IBSS). The ad hoc configuration is similar to a peer-to-peer office network in which no node necessarily functions as a server. As an ad hoc WLAN, laptops, desktops, and other 802.11 devices can share files without the use of a centralized AP.

Currently, WIDS technology (see Chapter 16) will not detect an authorized wireless device communicating peer-to-peer with an unauthorized wireless device. This scenario can create a bridge into the wired network by allowing an attacker to connect to a wireless device operating in ad hoc mode. The ad hoc mode allows a wireless device to relay traffic to the network and creates a number of potential attack scenarios.

Ensure that you disable the ad hoc mode for 802.11 unless your environment is such that you can tolerate the risk. You may find that some products do not allow disabling ad hoc mode, so use these devices with caution or use a different vendor. This is your warning: 802.11 ad hoc mode is exploitable. Users of hosts with ad hoc mode enabled may unintentionally allow users to inadvertently or maliciously connect to those systems.

In the future, a secure authentication framework will be developed for ad hoc mode (or peer-to-peer independent BSS). Then you can safely use ad hoc mode, but not before.

Disabling SNMP or Select Strong String

The Simple Network Management Protocol (SNMP) is widely used to monitor, manage, and control network devices using TCP/IP. Most network management systems (NMS), such as Cabletron Spectrum, HP OpenView, IBM Tivoli NetView, or Microsoft Network Monitor, can interface with SNMP. An SNMP browser or monitor is a very useful tool in most networks, more so in very complex ones. Part of the protocol includes a management information base, or MIB. Think of the MIB as a hierarchical tree similar to a DOS or UNIX directory structure with some predefined values (directory names) and some custom areas for vendors. Because this database includes all the manageable objects, MIB is the most basic element of network management. It is also a font of valuable and sensitive information.

So you have a great database of management information, but you need a method to access it. You can use an SNMP-enabled application to access the MIB. A basic one like `snmpwalk` can send SNMP commands and receive data. Another common application, a MIB Browser such as Solarwinds IP Network Browser (`http://solarwinds.net/Tools/Network_Discovery/IP_Network_Browser_SE/index.htm`), allows you to see all the branches and leaf nodes inside a MIB or group of MIBs. The full-blown NMS allows walking the tree and much, much more.

SNMP is not a very secure protocol, though it does have a minimal-security feature called SNMP community strings, which are like group-level passwords. They are shared passwords: Many people may know the community string. Since many people may know it, we lose accountability. Without another audit trail, you cannot tell whether Barry or Peter accessed the MIB.

There are three kinds of SNMP community strings: read-only, read-write, and trap. The *read-only* string allows you to make simple SNMP information requests. The *read-write* string also allows you to make information requests, but further allows you to change the value of a MIB object (assuming the object is updatable). The *trap* community string is rarely used, but allows network administrators to clump network entities together into groups, or communities. Then, you can configure the NMS to process traps received from one or more unique and specified communities. Since your NMS has its own way of limiting managed entities, this ability is usually redundant. To access one of the three functions, you need to know the associated community string. SNMP can trap authentication errors for reporting.

Obviously, there are lots of places you need and will use SNMP-enabled devices; however, a device on your perimeter is not one of them. You should change or, even better, disable SNMP community strings. Your access point comes with SNMP enabled. The vendor has to document clearly the read-only and read-write community strings in the manual that comes with your hardware. If it's available to you, then it's available to everyone. Using the default or weak strings, an unauthorized individual could use her SNMP agent to reconfigure your device in the same way as when she uses HTTP or Telnet.

So disable SNMP. If you insist on using it, reset the community strings to a hard-to-guess, non-default value. Some values to avoid are:

- Name of your company
- Name of your network manager or operator
- Location

- Device manufacturer
- Device name, part number (for example, WAP11), or type
- Name of your NMS
- SSID: either default or your actual network name
- Administrator password, or for that matter, any passcode that you use to access any system
- Any encryption key, especially your WEP key
- PUBLIC
- PRIVATE
- SNMP

You get the idea. Take your time and create a strong passcode. This is your first line of defense. When crackers own your network device, they own your network!

Don't forget to use access lists and firewalls to control access to ports 161 and 162.

If you are using a Cisco access point, then you have another decision. Cisco supports Cisco Discovery Protocol (CDP), which is a layer 2 (data link) protocol. CDP runs on all media supporting Subnetwork Access Protocol (SNAP). You can use CDP to obtain the addresses and other critical information, such as device type, patch level, and interfaces in the CISCO-CDP-MIB, from neighboring devices. Therefore, you should disable CDP as well. Your vendor may have another proprietary discovery protocol, and you should disable it to boot.

Security-conscious organizations will want to disable all configuration interfaces on access points and wireless bridges except for the serial console port or an RJ-45 jack.

Turning Down the Power

Now you may think that 1 Watt of power is not a lot, but it is. You might have a nightlight in your bathroom so when the little ones (or old ones) get up in the middle of the night, they can see their way around. A typical nightlight is about 7 Watts. It gives off a lot of light. However, nightlights generally are not focused, so the light they give off disperses. Were you to focus the light, you could see it up to 50 miles, or 83 kilometers, away, assuming there was no light pollution. Light does not penetrate walls, but you know that radio waves do. You have heard the hip-hop from your neighbor's, child's, or sibling's radio so you know that radio waves penetrate walls.

Your favorite FM radio station is somewhere between 88 MHz and 108 MHz. Long before the Nintendo generation, kids would lie in bed at night and listen to an FM channel broadcast hundreds of miles away. FM radio waves are longer than our 802.11 radio waves, so they travel farther, but they don't carry as much data.

The U.S. Department of the Navy uses ultra-low frequency to send messages to atomic submarines thousands of miles away under the polar ice cap. These messages use longer wavelengths and travel through water, not air, but the principle is the same.

One last example. You're sitting at a traffic light when your car starts to vibrate. Along comes a Honda Civic all tricked out with heavily tinted windows, lots of chrome, and the obligatory spoiler. You know the car we mean: the rolling boom box blasting music at you. As the car approaches, you hear the bass first. Why? Because base waves are the longest and reach you first. When the car pulls abreast of you, you get the full effect and hear other parts of the song. So, the longer waves of 802.11b and g (2.4 GHz) will travel farther than the shorter waves of 802.11a (5 GHz).

Assume that you can get a strong signal 100 meters from your access point. When you use a 16 dBi gain antenna, the range increases to 4 kilometers. Use a 20 dBi gain antenna and the range extends to 10 kilometers. Some companies have visual surveillance of their parking lots and feel that they would notice someone lurking in a car with a laptop. But the person doesn't have to sit in your parking lot when they can park 10 kilometers, or about 6.6 miles, from your office.

Because signals can bleed over beyond your perimeter, the first thing you need to do is use the power settings on your access point to do some cell sizing and cell shaping. Any access point that is not for the mass home market should provide the ability to tweak the power. Consider reducing the power of the access point to weaken the signal so that it spans a shorter distance.

Second, you need to test the signal strength outside your organization. If you find the signal is too strong, then you will need to introduce some loss. You can do this through the use of an attenuator. You can pick these up at any good electronics store or find them on the Web (for instance, at www.coaxicom.com/home.html).

And finally, you can use a technique known as RF signal shaping to *direction-alize* the RF signals emitted from your access point. You could switch from an omni-directional antenna to a semi-directional antenna to control the radiation pattern; otherwise, you may as well pull your UTP to the parking lot, as we have said numerous times.

Securing WAPs with a Subnet and a Firewall

Peter recently read another tome on wireless networks. This book focused on 802.11 and did not touch WPANs, WMANs, or WWANs. All in all, the book was good; however, one section was quite shocking. After seeing this shocking section, we checked other books and found that they had a similar problem. Look at Figure 18-1 and see if you can find the problem.

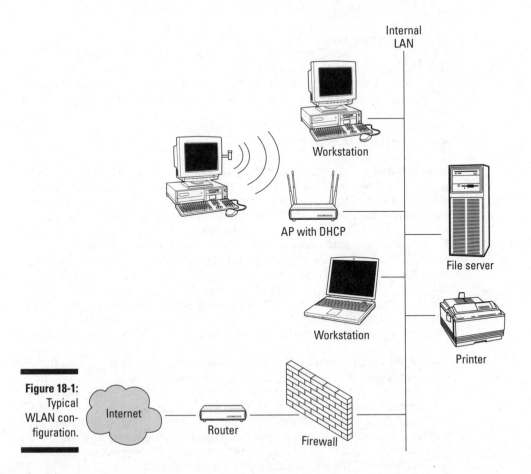

Figure 18-1:
Typical
WLAN con-
figuration.

That's right, the access point is behind the firewall. This is not unusual. In the not-so-distant past, we saw the same problem with dial-up. Individuals asked for and got an additional analog line in their cell, cubicle, or office, and used it to do remote control. They used their modem and software like PCAnywhere, VNC, ReachOut, Relay Gold, Timbuktu, Carbon Copy, or a similar product to dial in and access their desktop to get to the internal network. Sometimes, these individuals accessed their e-mail; other times, they used the connection to get to the Internet without paying an ISP. Regardless of the reason, they were exposing their organization's network. We worked hard to get them to centralize the modems and use remote node rather than remote control. Now, we have set up a different but analogous situation, and a potentially more dangerous one. The network or security administrator is focusing on the front door (the firewall) as everybody is running in and out the backdoor (the access point). To illustrate, think of your access point as the equivalent to connecting some cable to your hub and pulling it outside to your parking lot and letting anyone use it. You cannot (and must not) trust your wireless network. It is a public medium just as the air it uses is public (for all intents and purposes).

Most network administrators would never think to connect a server directly to the Internet. Instead, the administrator would put it behind a firewall and hide the device from the outside. However, when a wireless workstation connects directly to your network behind your firewall, you have bypassed your firewall and negated its effectiveness. This is as dangerous as connecting that server directly to the Internet. What happens when the owner of the wireless workstation fails to install critical patches or virus detection/prevention software? You have put everybody at risk.

You must protect yourself: You need to take steps to ensure that you wall off wireless workstations from wired workstations. If your wireless workstations don't need access to the rest of your network, this is simple. Put up a big wall. Create an air gap. But the reason we use wireless is to extend our network, so we need a method to handle the access securely. The method we use is called a wireless demilitarized zone or WDMZ. If you are familiar with the TV show *M*A*S*H,* or have followed the "Axis (or is it Access in our case) of Evil" diatribe, then you probably know about a DMZ. The aforementioned demilitarized zone is a strip of land created by the truce at the end of the Korean War. This DMZ is a "no man's land" between North and South Korea: Show up in the DMZ and both sides will take a shot at you. In our terminology, it's not so severe, but the DMZ is definitely in the wild, wild west, where different laws apply.

We suggest that you add a DMZ or screened wireless network between the internal network (your intranet) and the external network (the Internet). In this DMZ, you place your authentication server, Web server, and external DNS server. You use an authentication server to regulate traffic between the untrusted network (the Internet) and the trusted network (the intranet). With wireless, you segregate your access points and have a trusted way into the internal network. In security terminology, this is called *compartmentalization,* in which wireless stations are segregated onto one or more separate network segments, and direct communication with other devices on the wired portion of the network is prevented without authentication. By compartmentalizing, we can isolate risks and apply controls to mitigate or eliminate the risks.

The simplest WDMZ configuration is shown in Figure 18-2. Notice that we used an authentication server to connect the wireless network to the internal network. This server acts as a proxy and handles the inbound request and hides the internal network from the outside. We also have a dedicated DHCP (dynamic host configuration protocol) server to serve up dynamic IP addresses for the wireless portion.

The configuration in Figure 18-2 is an improvement over Figure 18-1, but it's not ideal. You still bypass the rules in your firewall. A goal for network security is to minimize the number of entry points into your organization. This configuration doesn't do that. In Figure 18-3, we move the WDMZ to the firewall itself. This is a comparable situation to what we did with our dial-up servers. Note that clients still need to authenticate themselves at layer 2, but we can use the firewall to block traffic or require user authentication. In Chapter 11, we introduce you to network security features such as WEP and EAP. This is where they all come together.

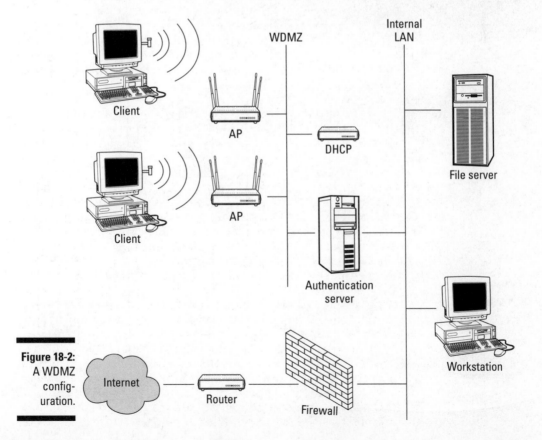

Figure 18-2:
A WDMZ config-
uration.

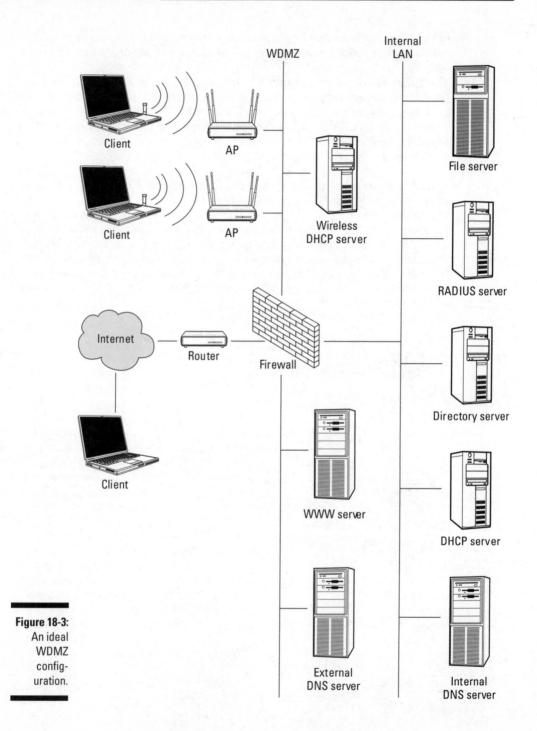

Figure 18-3:
An ideal
WDMZ
config-
uration.

Oh, and one last network issue. Always connect access points to switches, not hubs. Hubs, like access points, are broadcast devices, so everyone hears everything. The hub will send out every packet the hub receives, including those from the wired segment.

Using a WIDS

The threats to your WLANs are many and potentially devastating. Security issues ranging from misconfigured wireless access points to session hijacking to Denial of Service (DoS) can plague a WLAN. Not only are wireless networks susceptible to TCP/IP-based attacks, but they are also subject to a wide array of 802.11-specific threats. To help you defend and detect against these potential threats, you need to employ a security solution that includes a *wireless intrusion detection system* (WIDS). Even organizations without a WLAN are at risk from wireless threats and should consider a WIDS solution.

Chapter 16 deals with using a WIDS and a NIDS together in your organization.

Disabling Wired Access from Public Areas

Now we probably shouldn't complain, because sometimes it makes our life easier. But you really should disable wired access from public areas. Wait, isn't this book about wireless? Well, yes it is. But this book also spends a lot of time talking about wireless access points in infrastructure mode. Infrastructure mode infers that you connect the AP to a wired network. Above, we talk about rogue access points. Well, here is the nexus: rogue WAPs and wired networks. While someone is visiting your office, perhaps cooling their jets in a waiting area, they could see a faceplate. They could have the wireless bridge mentioned above. They could plug it into the RJ-45 jack, and, bingo, they can walk outside your building, but continue to have access to the wired network. If they are bad, they use Ethereal to look for usernames and cleartext passwords. Maybe they use nmap to scan your network looking for problematic applications. Worse still, they leave it as an open access point and all your neighbors take advantage of the bandwidth to download pornography or attack another company. The liability will rest with your organization. Protect yourself and your assets. Ensure you know every entry point into your organization.

Hardening the Access Point and Clients

Access point hardening is the step-by-step process of securely configuring a system to protect it against unauthorized access, while also taking steps to make the device more reliable. Generally, hardening ensures the system is both more secure and more reliable.

You will find that access point hardening is necessary because most vendors design access points to be easy to use, rather than secure, out of the box. You can, however, enable the built-in features to make the AP more secure. Following is a simple list to ensure your access point is hardened:

- ✔ Ensure you have the latest firmware for your system.
- ✔ Disable SNMP or use strong community strings.
- ✔ Change the default administrative passcode to a strong one that is difficult to guess.
- ✔ Filter traffic to the administrative port or protocol used to manage the device itself.
- ✔ Use the strongest encryption available to you.
- ✔ Remove unnecessary features or protocols.
- ✔ Use enhanced authentication such as EAP (see Chapter 11).
- ✔ Place the access point in a physically secure location to prevent tampering or theft.
- ✔ Protect your access point against natural phenomena such as lightning and strong winds.
- ✔ Periodically test for jamming attempts using a spectrum analyzer.

Wow, another list of ten!

Access point hardening is only part of a secure network environment. You also need to establish strong policies and communicate them to your fellow workers.

After hardening your access point, you need to do the same to your clients. You need to ensure that they all have a current version of your favorite anti-virus software as well as a personal firewall, such as ZoneAlarm (www.zone labs.com/store/content/home.jsp) or BlackICE (http://blackice.iss. net). Both are good products.

Also, remember that the keys are stored on the wireless adapter, so protect the PC card or USB device.

Chapter 19

Ten Ways Wireless Is Used in Business

There is absolutely no reason to set up a wireless network if you can't make good use of it. So we decided to provide you with a chapter to get your right brain thinking about how you may use wireless in your organization. Look over the ten ways here and adapt, adopt, extrapolate, or discard them to come up with your own top-ten list.

Attending Meetings with Tablet PCs

Not all Tablet PCs come with Centrino chipsets, but enough do to make this scenario real. It's not uncommon to attend a meeting in which someone has an HP Compaq Tablet PC TC1100. The TC1100 is a lightweight (3 pounds), high-performance PC with an innovative form factor.

Most tablets have a keyboard that the manufacturer hinges to the screen. You then bend the keyboard back and around and it lays flat under the screen when you want to use the PC as a tablet. The TC1100, on the other hand, has a detachable keyboard, leaving the screen as the tablet. The ports and slots are in the tablet, not in the keyboard part as in a laptop. The removable keyboard provides the ability to change from tablet mode to notebook mode with the flip of a switch. You can use the tablet in portrait or landscape mode, whichever suits your style.

HP provides Microsoft Windows XP Tablet PC Edition for Mobile Computing with the TC1100. Since this operating system is a subset of Windows XP, it is fully 802.11-aware. The TC1100 also has Centrino, which means it has onboard 802.11b support.

The ultra-mobile Compaq tablet is a little different than other tablets. It is the convergence of handwriting and computing. For one thing, the tablet adapts to the way you work. You have the simplicity of pen and paper, and you can write the way you naturally write. The OS learns how you write. You don't have to learn how to write script the way the OS wants you to. (Peter once had a Newton. He felt like he was back in kindergarten as the Newton OS constantly criticized his writing.) As you take notes during the meeting, you can store the information as *digital ink*. Since it's digital, it's like paper, only better; so you can change the ink color, use highlighters, insert space, erase text, italicize words, draw pictures, and doodle to your heart's content. Before you leave the meeting, you can send minutes to everybody who attended, so that they will have the minutes the minute they get back to their cubicle or office. Now that's efficiency. And very, very scary.

For those who like to stay informed when in meetings, you can do online searches of the Web in real-time. When your boss tells you she is sending you to the Republic of Azerbaijan to check the oil reserves, you can check it out at www.cia.gov/cia/publications/factbook/geos/aj.html#Econ. When she tells you that you can spend a weekend at the Apsheron Peninsula on the Caspian Sea as a reward, you can politely decline on the spot.

Think of the endless possibilities as you communicate quickly, discretely, and discreetly with others using digital ink. Dilbert and you can start a little cabal against the pointy-haired guy without even saying a word.

Of course, the TC1100 has a long battery life for those oh-so-long meetings. Now you and your friends can play buzzword bingo while biding your time. You won't have to print your card before the meeting. Just go to http://isd.usc.edu/~karl/Bingo/bbbingo.html, send your friends in the meeting an instant message, and start playing. You can use the highlighter feature and mark the squares as you hear the buzzwords. We suggest that, rather than yelling out "Bingo," you send another IM — you don't want to wake the people around you. Isn't technology wonderful?

Getting Your E-mail As You Wander the Building

If you follow the advice of this book, you will set up a wireless demilitarized zone (WDMZ) to alert you of a potential attack. (See Chapter 16 for further information on WDMZ.) You can use a specialized hardware device or you can use a wireless network analyzer. It doesn't matter, as long as it can generate legitimate alerts. When a potential attack or problem occurs, you'll want to send an alert.

Presently, most administrators send an alert to a text pager, but you could send an e-mail to your Sidekick, Blackberry, or Tablet PC. This way, you can remain on the job even while hanging around the water cooler or while on your way to have a smoke. You could also execute a program to take action when necessary. What's more, you could forward the message to someone who cares. Hey, it's 4:00 a.m. Need we say more?

Getting Corporate Access in the Lunchroom

Many organizations have policies that basically forbid the use of the Internet during working hours, breaks excluded. These same organizations don't mind employees accessing the Internet for legal purposes on their own time.

Also, many organizations provide the menu for the cafeteria online so that employees can judge whether it is worth making the long trek to the cafeteria for lunch.

So, providing access to the Internet and other network resources from the lunchroom seems a perfect marriage. Why not provide an access point and allow employees to access, or even order from, the cafeteria menu, download their e-mail, check CNN or ESPN, or look up other trivia?

Setting Up Wireless Conference Rooms

When setting up a wireless conference room, make sure that the conference room gets a signal. There is more than one story out there about a conference room with no signal. One organization used RF prediction software and placed access points in strategic locations throughout the building. Everything worked reasonably well, except they couldn't get a decent signal in the conference room. They could not understand why; the access point was placed right outside the room. Eventually, they looked at the original architect's drawings. They found out that they had taken over the former premises of a medical clinic. Even though the clinic was not physically collocated with the hospital, it had an X-ray machine. The machine was previously housed in the room that the new tenant thought would make a good conference room. Unfortunately, only then did they learn that the walls were lead-lined. Lead stops signals dead. Just ask the CIA.

Another signal killer is interior walls. Plaster with wire mesh and rebar concrete act as a Faraday cage (`http://en.wikipedia.org/wiki/Faraday_cage`) in practice and will impact the use of wireless networks inside your building. A *Faraday cage* is an enclosure with no holes, slits, windows, or doors and is made of a perfectly conducting material, such as the wire mesh wall. When there are no electric fields produced within the Faraday cage (for example, your boardroom), the mesh becomes an electromagnetic shield. This means that no electrical energy sources can penetrate the conducting enclosure. So when you don't have a signal inside the room, none can penetrate. This is why RF prediction software is useful, but a site survey is required. You never know with RF; for all intents and purposes, it is unpredictable.

There is science behind the Faraday cage. Usually a complete conductive shell, for example, your wire mesh wall, collects stray charges and, because like charges repel, stores them on the outside surface. They are stored farther apart than on the inside. The electric fields generated by these charges then cancel each other out on the inside of the cage. This prevents the movement of the electric charge into the enclosed area. Governments, scientists, and manufacturers often use a Faraday cage to protect sensitive radio equipment.

We'll assume that you have ensured that your walls aren't lead-lined and in no other way kill your signal. Now it is just a simple matter of putting in a captive portal, when it is an open conference room. (See the discussion about captive portals that follows.) Make sure that your captive portal works. Peter was in a wonderful educational facility in Chicago. The facility was clean, bright, modern, and very state-of-the-art. At the reception desk was a sign stating that wireless was available in the individual conference rooms. He inquired and was told that he would need a username and password, which they provided. Peter fired up his browser and had access to his homepage. Wait a minute — shouldn't there be a captive portal? He reported it to the concierge who replied, "Oh, they were supposed to fix that. I guess they didn't. Go ahead and use it."

We also have seen instances in which the wireless network planner did not account for people in the conference room. You may get a strong signal when the room is empty, but add several dozen people and you may find that the signal is weak. Humans are not conductors; they will absorb RF signals.

Querying Your Corporate Database

You are walking around the warehouse, and you stop by the bin where they store wireless networking PC cards. You can eyeball the bin and make a guesstimate of the quantity. But wouldn't you really want to know how many there really are in the bin? Well, with a wireless connection, you could use your PDA or laptop to send a query to your backend database management system and find out.

You are sitting in a sales meeting in the boardroom and someone asks for the sales projections for the month. You could take out your laptop, find a face-plate, plug it in, and try to access your sales and marketing data. Problem is, it might not work. Your organization might not make the RJ-45 jack live unless requested. Or, your organization might firewall off the boardroom from the backend databases for security's sake. (This happened many times to Peter.) Now, wouldn't it be easier to take out your PDA, connect to the access point, do the query, and tell everyone the result?

Keeping in Touch at the Airport

Not long ago Peter was in transit at the Pittsburgh International Airport (www. pitairport.com). When you have to cool your jets in an airport, the Pitt airport is one of the best places to do it. Pittsburgh International Airport is world-renowned and ranked among the top airports by *Official Airline Guide,* J.D. Edwards, *Conde Nast Traveler* magazine, Travelocity, and e-Travel.

After you get through security, you will find that the airside terminal is a lot like a shopping mall. In fact, Allegheny County Airport Authority even calls it an Airmall. There is a center core with spokes radiating out. The spokes have the gates. The core has the shops and the food zoo. When Peter was there, there were four pilots sitting at a table in the food zoo with their laptops touching each other, staring intently at their individual screens. Now, this may qualify as getting corporate access in the lunchroom, but we consider it different.

Now, many airports have high-speed Internet access at for-profit kiosks (for example, www.powerinternetkiosk.com or www.globalinternet terminals.com/kioskgle.html), but Pitt is different, because it offers free-of-charge access to wireless access points. The airport authority installed five hotspots at the airport, and they advertise that passengers can connect in the food court areas in the Airside Terminal's Center Core and A and B concourses. So you can sit in the food zoo with the pilots, or you can use it from most gates. When Peter tried to use it at his gate, he had to reorient the antenna to get a strong enough signal to send and receive e-mail. However, in the food zoo, it was smoking! You will find that you will experience Internet connections at speeds as fast or faster than many broadband services.

It's simple to use. You just turn on your computer and use basic default settings. Your adapter will find the wireless connection automatically.

Pittsburgh International Airport boasts that it is among only a few airports in the country equipped for wireless communications, and one of the only ones that does not charge a fee. This might not always be the case. On the same trip as the stopover in Pittsburgh, Peter was in the Williamsport-Lycoming Airport. If you are a sports fan, you might recognize Williamsport, Pennsylvania, as the home of the Little League World Series. He had some time to kill, so he took out his laptop and "discovered" there was an access point with an SSID of Linksys with no encryption. To this day, Peter is not sure whether this is a service like Pittsburgh's service. If it is, the airport authority should advertise it. If it isn't, they should protect the access point.

To save time and to determine whether you can even get a connection, you can use a device like the Kensington Wi-Fi Finder (www.kensington.com/html/3720.html). The Wi-Fi Finder is a keychain-sized device (2.95"L x 0.39"H x 2.17"W) that detects 802.11b/g signals up to 200 feet away. You don't have to take out your laptop and use your wireless utility. Instead, you just press a button. There are three LEDs that indicate signal strength. The more lights, the stronger the signal.

Maintaining a Presence While Having Coffee

You may have seen the Bud Light commercial in which an employee leaves a steaming cup of coffee on his desk. The coffee cup generates the steam; it doesn't really come from the coffee itself.

His boss comes by the office, sees the steaming coffee, assumes the employee is in, and remarks that he is in early, while in reality the employee is still in his cozy bed. During the day, other employees come by the office, see the coffee and assume that the employee just stepped away for a minute and will return shortly for his coffee. At the end of the day, the boss comes by the office again on his way out, sees the steaming coffee on the desk and praises the industriousness of the employee. Meanwhile, the employee is whooping it up at the bar.

Clearly, we're not with the Bud Light Institute, but we can help you out, too. With wireless local and wide area networking, you can have your coffee while making your boss think you are slaving away at your desk. You can visit your local Starbucks (`www.starbucks.com/retail/wireless.asp`), where they provide a hotspot, settle into one of their comfy chairs, and continue to access your e-mail and look up things on the Internet. You will impress your boss when you answer your e-mail after hours. Your boss doesn't need to know you are chilling with an iced caramel macchiato and nibbling on a chocolate biscotti. That will be our secret.

Starbucks uses T-Mobile. You can get a one-day pass from `https://selfcare.hotspot.t-mobile.com/accountcreate/ExternalSetPromotionCode.do?promo=STDAYPASS0802`, but you will have to enroll to get it. As a cell phone company, T-Mobile got into commercial hotspots in a big way, first by acquiring MobileStar in 2002. T-Mobile has more hotspots than any of its competitors. They have about 5,000 hotspots in every state except Alabama, Alaska, Arkansas, Montana, North Dakota, South Carolina, West Virginia, and Wyoming. In addition to Starbucks coffeehouses, T-Mobile has hotspots in Borders Books & Music stores, Kinko's, airports, and the airline clubs of American, Delta, United, and US Airways.

To use T-Mobile at Starbucks, follow these simple steps:

1. **Using the configuration utility for your wireless adapter, create a new profile.**

   See Chapter 4 if you need help with this step.

2. **Set the SSID to T-Mobile.**

   If you are not sure what the SSID is, use the network discovery feature of your client configuration utility.

3. **Disable encryption.**

4. **You will need to ensure you renew the IP address, so select that option.**

   When you sign on to T-Mobile, your adapter will get a dynamic IP address.

5. **Save the profile. If your client doesn't see the access point, then ensure you did steps 1 through 4 correctly.**

6. **Open your browser.**

   When your browser tries to open your default home page, T-Mobile will capture it and display their login Web page.

7. **If you pre-registered and have an account, enter the username and password in the appropriate box. If you don't have an account, you can click New Users Click Here to get to the sign-up page. Enter the promotion code for Starbucks and click the SIGN UP button to see the payment options.**

8. **When you are finished, click the logoff link so that T-Mobile stops charging you for access.**

You should make yourself aware of the additional security risks of using a public hotspot. Unlike your internal networks, in which you have trusted and untrusted users, you now have those who can and those who cannot join the network. Those who can join the network at Starbucks are not necessarily trusted. In fact, you should take the security stance that they are not trusted and mean you harm. So, you should take care to protect yourself and your data. If accessing your organization's network from the wireless hotspot, you may want to use a VPN client. If you haven't read Chapter 12 on VPNs, you may want to look at it now.

Using Bluetooth Phones in Your Car

Police say that using a cell phone while driving a car can equate to .08 milligrams of alcohol — that is, legally drunk in most places. Put a small dog in the person's lap and they might as well be falling down drunk. Passengers, on the other hand, can use the cell phone as they please. But have you ever noticed that the person on the phone when the car has passengers is always the person who is driving?

Anyway, with a Bluetooth-enabled phone or a phone with a wireless PC card that can communicate using the cellular network, you can connect your laptop (which is plugged into the cigarette lighter using an inverter) or PDA to the Internet. If you are using Bluetooth and you see someone slipstreaming you, you might be providing him access to the Internet as well. You don't believe me? Convoys to geek-fests in the past have created CANs — and we don't mean Campbells soup cans connected by waxed string. We mean *Car Area Networks*. As cell networks get more bandwidth (look for 2.5 or 3G phones), you can expect more CANs to pop up. (Or is that, pop open?)

You can find a list of Bluetooth-enabled phones at `www.bluetooth.com/products`.

Auto manufacturers Acura, Audi, BMW, DaimlerChrylser, General Motors, Lexus, and Lincoln have shipped cars with Bluetooth technology, giving drivers a true hands-free calling experience. These carmakers provide the hands-free system using your car's stereo system as the microphone and speaker.

But it's not just Bluetooth-enabled phones. Citroën, the French car manufacturer, demonstrated the use of Wi-Fi in a C3 Pluriel (`www.gizmo.com.au/public/News/news.asp?articleid=1984`) to provide low-cost Internet access while on the road. The Pluriel is equipped with both Wi-Fi and a 3G mobile phone. This combination allows the car to become a Wi-Fi hotspot. A Toronto company, Baka Wireless (`www.baka.ca`), has a Smart car hotspot that they drive around to trade shows. Florida's MESH Networks (`www.meshnetworks.com/pages/technology/intro_technology.htm`) is fast pulling away from the competition; it can wire (or is that unwire?) almost any moving vehicle to get 2 Mbps. Have you driven a hotspot lately?

The next time someone asks whether you are driving a van, don't be insulted. They may mean a Vehicle Area Network!

If you use an antenna for this network, be careful you don't cut or scratch yourself; you could get van-aerial disease!

Accessing a Wireless Network in Your Hotel

In this day of cookie-cutter hotels, network access is an important differentiator for hotels wanting to attract business clientele. To this end, many hotels have signed exclusive deals with hotspot providers. Hotels generally want to get involved in the installation and operation of any wireless network, so they are most likely more secure than a coffee shop on your corner.

If you are ever in Jukkasjärvi, Sweden, 125 miles north of the Arctic Circle, you can visit the Ice Hotel (`www.icehotel.com/english/index2.htm`). You may have seen the TV ads for the Ice Hotel promoting a prominent credit card company. Opened in 1992, the owners rebuild the entire hotel every year, including the bar, out of ice. Now, you really can have your Aquavit on the rocks here. They needed to install a network in the hotel, especially for the bar. Writing up orders when the ink in your pen doesn't flow well is a challenge. But can you imagine drilling holes in the beautiful pristine ice to pull cables? Sacrilege. So they decided to install a wireless network instead. You may have had no problem when you installed your wireless access point, but try doing it in an icehouse. The radiated power heats the ice. Ice doesn't transmit signal very well. Your usual problems. But, they made it work and are planning to expand the wireless to other areas.

Chances are your hotel didn't have as difficult a time installing a wireless network. Before setting out on a trip, you can find hotel hotspots using the following Web pages:

- ✔ EZGoal Wi-Fi HotSpots: `www.ezgoal.com/hotspots/wireless/s.asp?qu=Hotel`
- ✔ HighTech Traveler: `www.hightechtraveler.com/cgi-bin/page.cgi?d=1`
- ✔ HotSpotList: `www.wi-fihotspotlist.com`
- ✔ Hotspot Locations-The Wireless Directory: `www.hotspot-locations.com`
- ✔ Jiwire: `www.jiwire.com`
- ✔ The Wireless Node Database Project: `www.nodedb.com`
- ✔ Wi-Fi Free Spot: `www.wififreespot.com`
- ✔ Wi-Fi Networking News: `http://wifinetnews.com`
- ✔ WiFi 411: `www.wifi411.com/search/index.php`
- ✔ WiFinder: `www.wifinder.com`

Unfortunately, you need to be online to read these lists. Aye, there's the rub. That is why a client like Boingo (see Chapter 6) is so useful. Boingo has included the connection information into the client. So, when you are visiting the world-famous Hershey Chocolate World and HERSHEYPARK amusement park, you find that there's a hotspot at the Hershey Lodge and Convention Center on West Chocolate Avenue. (We kid you not; we can't make this stuff up.)

Presently, Wayport (`www.wayport.com`) is the number-one provider to the hotel market. Wayport uses a captive portal for security. Start up your browser and try to visit any page and the captive portal redirects you to a Web page where it will prompt you for some additional information, such as agreement to a room charge, a valid account, or credit card information. The captive portal does the following:

- ✔ The portal acts as a firewall between the wired and wireless networks. It may consult with a backend authentication system before relaxing its control.
- ✔ It identifies and authenticates paying or authorized clients.

✔ It manages network traffic priority according to a scheme. The node owner usually has the highest priority, network members have lower priority, and nonmembers have little or no priority.

✔ It sets bit rate limits on traffic based on the type of client. Again, the node owner usually has the highest limit, network members have a lower limit, and nonmembers have little or no limit.

✔ It overrides the firmware on the client adapter and makes the client adapter adapt.

If you want to build a captive portal for your organization, check out Cheshire (http://nocat.net/download/Cheshire), NoCatAuth (http://nocat.net), NoCatSplash (http://nocat.net/download/NoCatSplash), OpenAP (http://opensource.instant802.com) or Sputnik (www.sputnik.com). You could also use a proxy server such as Squid (www.squid-cache.org) to control wireless access.

Back to the hotel. Connecting to Wayport is as easy as following these steps:

1. **Using the configuration utility for your wireless adapter, create a new profile.**

   The hotspot should look wide-open to your configuration utility. See Chapter 4 if you need help with this step.

2. **Set the SSID to Wayport_Access (with capital W and A and an underscore between the two words).**

   If you are not sure what the SSID is, use the network discovery feature of your client configuration utility.

3. **Disable encryption.**

4. **You will need to ensure you renew the IP address, so select that option.**

   When you sign on to Wayport, your adapter will get a dynamic IP address.

5. **Save the profile. If your client doesn't see the access point, then ensure you did steps 1 through 4 correctly.**

6. **Open your browser.**

   When your browser tries to open your default home page, Wayport will capture it and display their portal entry.

7. **If you pre-registered and have an account, click the Use Wayport Membership button and enter the username and password in the appropriate box. If you don't have an account, you can click either the Purchase a Connection or the Use a Coupon button.**

Click the former, and you will need to read the information on the Web page and click CLICK HERE TO CONTINUE. If you select the latter, you need to enter the coupon code before continuing.

8. **Enter the promotion code for your hotel.**

   In one way or another, you will need to authenticate yourself to the portal. And money talks here as it does most places.

9. **After you log in, Wayport displays a Web page that requires you to click GO TO REQUESTED WEBSITE to get the page you requested before the redirection.**

10. **When you are finished, close your browser.**

Using the Phone to Check Your Stocks

Using a service like Stock Smart (`www.stocksmart.com/ss/html/hpcellphone.html`), you can access constantly updated stock quotes, your portfolio, up-to-the-minute indexes, current IPOs, and breaking NASDAQ news from your cell phone. All you need is a Web-enabled cell phone with a mini-browser. Of course, you also need some money to pay for the service.

To use Stock Smart, just enter the address WAP.STOCKSMART.COM. If you are pre-registered, highlight Login and press OK. Enter your username and password on the appropriate screens. You are connected and can access stock information.

Alert Guru (`www.alertguru.com`) and TickTac (`http://eng.ticktac.com`) also will provide real-time quotes to your cell phone (or by e-mail).

In addition, many publicly traded companies prominently display a ticker tape of the company's stock price right on the homepage. So why not allow employees access to that same information wirelessly from the lunchroom or conference room?

Part VI
Appendixes

The 5th Wave   By Rich Tennant

"Frankly, the idea of an entirely wireless future
scares me to death."

In this part . . .

This part provides grist for your insatiable appetite. You get all kinds of fascinating information. We included some very detailed wireless information for the technically inclined, along with trade associations you can contact for additional information and wireless industry standards you can follow.

Appendix A

Industry Trade Associations

● ●

In This Chapter

▶ Government organizations

▶ International and national standards organizations

▶ Wireless-related organizations and associations

▶ Local wireless groups

▶ Other industry associations

● ●

*W*ireless networking is evolving extremely quickly. To keep your company and yourself current, you need to keep up-to-date on developing
standards and products. This book is your launching pad, but you'll need to
do further research (or buy the 2nd edition of this book). We have listed
some of the organizations and their missions that can help.

Government Organizations

Individual governments control the spectrum within their jurisdictions.
You will need to familiarize yourself with the laws and regulations in your
jurisdiction.

✔ **Federal Communications Commission (FCC):** In the United States, the
FCC controls the airwaves. Federal law governs the use of Wi-Fi. The FCC
administers "Part 15 of Title 47 of the Code of Federal Regulations." Part 15
embodies the rules and regulations for the unlicensed use of the radio
spectrum and applies to microwave ovens, shortwave diathermy medical devices, cordless phones, low power "walkie-talkie" 2-way radios,
and 802.11 wireless equipment. You can find the FCC at www.fcc.gov.

✔ **Canadian Radio-television and Telecommunications Commission
(CRTC):** In Canada, the CRTC controls the airwaves. This department
administers the Broadcasting Act, including the unlicensed radio spectrum. You can find the CRTC at www.crtc.gc.ca/eng/welcome.htm.

✔ **Office of Communications (Ofcom):** The Office of Communications, started in late 2003, is the regulator for the United Kingdom communications industries, with responsibilities across television, radio, telecommunications, and wireless communications services. Ofcom replaced the Radio Authority. You can find them at www.ofcom.org.uk.

International Standards Organizations

There also are international standards groups who cut across boundaries, just as radio waves can. The following groups come into play when working with either WPAN, WLAN, WMAN, or WWAN equipment or looking for global standards.

✔ **3G Americas:** 3G Americas is a wireless organization formed to unify the Americas through wireless technology and to work collaboratively on furthering the definition and standardization of prioritized requirements with global and regional standards bodies for the deployment of commercial products and services in a timely fashion. This includes GPRS (General Packet Radio Service), GSM, EDGE, and UMTS (WCDMA). You can find 3G Americas at www.3gamericas.org/English/index.cfm.

✔ **European Telecommunications Standards Institute (ETSI):** ETSI has a similar role to the IEEE (see the next bullet), but it does it in Europe, as the name implies. ETSI established the HiPerLAN/1 and HiPerLAN/2 standards, the latter competing with 802.11a. ETSI and IEEE have discussed unifying on certain wireless technologies. They started the "5UP" initiative. No, it is not a soft drink competing with Code Red or Red Bull, but rather the "5 GHz Unified Protocol."

In Appendix B, you see a brief note on 802.11h, which is the IEEE's attempt to interoperate with the HiPerLAN/2 standard. HiPerLAN/1 supported rates up to 24 Mbps with a range of 45.7 meters using DSSS and the lower to middle U-NII bands. HiPerLAN/2 supports data rates up to 54 Mbps and uses all 3 U-NII bands. (U-NII stands for *Unlicensed National Information Infrastructure.* See Appendix C.) In the spring of 1997, the ETSI established BRAN (Broadband Radio Access Networks) as the successor to the former Sub-Technical Committee that developed the HiPerLAN/1 specifications. ETSI BRAN currently produces specifications for three major Standard Areas:

• HiPerLAN/2, a mobile broadband short-range access network

• HIPERACCESS, a fixed wireless broadband access network

• HIPERMAN, a fixed wireless access network operating below 11 GHz

You can find the ETSI at www.etsi.org.

✔ **Institute of Electrical and Electronics Engineers (IEEE):** In Appendix B, we cover the pertinent wireless standards. These standards are all IEEE standards. The IEEE leads the way in developing open standards for Wireless Local Area Networks (Wireless LANs), Wireless Personal Area Networks (Wireless PANs), and Wireless Metropolitan Area Networks (Wireless MANs). You can compare the 802.11 wireless standards for "over the air" to the 802.3 Ethernet standards for "over the wire." You can find the IEEE at `www.ieee.org`.

✔ **International Telecommunications Union (ITU):** The ITU Radiocommunication sector (ITU-R) has a role in the management of the radio-frequency spectrum and satellite orbits, in-demand finite natural resources such as satellites, and those communication services that ensure safety of life at sea and in the skies. You can find the ITU at `www.itu.int`.

✔ **Telecommunications Industry Association (TIA):** The TIA has standards for IMT-2000, also known as the Third Generation (3G) Mobile Systems. The goal of 3G is to make anywhere, anytime communications a reality. IMT-2000 is the result of collaboration among many entities inside the ITU (ITU-R and ITU-T) and outside the ITU (3GPP, 3GPP2, TIA, etc.). You can find the TIA at `www.tiaonline.org/index.cfm`.

Wireless-Related Organizations and Associations

No, not all of wireless is controlled by Big Brother; that's just paranoia. There are organizations and associations that aren't controlled by the government, but by interest groups. Some groups focus on a particular standard, and others focus on providing free access to the masses.

✔ **The 3rd Generation Partnership Project (3GPP):** The 3GPP, located in Sophia Antipolis, France, is a collaborative agreement between Standards Development Organizations and other related bodies for the production of a complete set of globally applicable technical specifications and reports for a 3G System. This organization and its various subgroups are working on setting WWAN and phone standards. You can find the 3GPP at `www.3gpp.org`.

✔ **Bluetooth Special Interest Group (SIG):** SIG is the trade association responsible for Bluetooth wireless technology. They have working groups for cars, audio/video, HCI, HID, ISDN, local positioning, PAN, printing, radio, still image, and UDI. In addition, there are expert groups for security and short-range financial services. Something for everyone! You can find Bluetooth at `www.bluetooth.com`, the official Bluetooth Web site, and `www.bluetooth.org`, the official Bluetooth membership site.

✔ **CommunityWireless:** Do you want to start your own free metro wireless data network? If so, then you might want to consult CommunityWireless. CommunityWireless is an umbrella organization that represents the emerging community networks and their needs. This organization aims to promote the use of off-the-shelf and license-free Wireless LAN technology to provide free wireless networks for communities. CommunityWireless provides funding and support to Wiana (covered later in this list) from the wireless communities it represents. You can find CommunityWireless at `www.communitywireless.org`.

✔ **Freenetworks.org:** The Freenetworks.org mission statement is "Free Networks.org is a voluntary cooperative association dedicated to education, collaboration, and advocacy of the creation of free digital network infrastructures." This ethos is similar to that held by the original developers of the Internet, and we know what happened there! If the big network and Internet service providers don't step in and try to kill the free networks, capitalism probably will. You can find out more about this organization at `www.freenetworks.org`.

✔ **Infrared Data Association (IrDA):** This organization advertises itself as "The Trade Association for Defining Infrared Standards." Founded in June of 1993, the IrDA is a member-funded association whose charter is "to create an interoperable, low-cost, low-power, half-duplex, serial data interconnection standard that supports a walk-up point-to-point user model that is adaptable to a wide range of computer devices." They are most noted for in-room line-of-sight networks, handheld computers, and personal digital assistants. One of our favorite restaurants in San Francisco (and now Beverly Hills), The Stinking Rose (`www.thestinkingrose.com`) uses an infrared network for order taking. You can find the IrDA at `www.irda.org`.

✔ **Orthogonal Frequency Division Multiplexing (OFDM) Forum:** The OFDM Forum is a voluntary association of hardware manufacturers, software developers, and other users of OFDM technology in wireless applications. Started in December 1999, the OFDM Forum has a mandate to foster a single, compatible OFDM standard. OFDM is a cornerstone technology for IEEE 802.11a, IEEE 802.11g, and ETSI BRAN (Broadband Radio Access Networks, discussed previously). You can find the OFDM Forum at `www.ofdm-forum.com`.

✔ **The Wireless Internet Assigned Numbers Authority (Wiana):** Wiana is a community-driven organization funded through CommunityWireless. The organization wants to ensure the most accessible, stable, secure, and cost-effective (that is, free) wireless address management. Wiana feels that the existing Internet registries cannot serve end users who now own their "last mile," so they have stepped into the void. They provide the following:

- IP addresses in the 1.x.x.x range for wireless devices, free of charge

- Certificates for each address to ensure full authentication using Wiana cryptographic protocols

- A centralized abuse-handling and blocking structure

- A mechanism for ensuring that their network policies are upheld

You can find Wiana at `www.wiana.org`.

✔ **The Wireless LAN Association (WLANA):** WLANA is a nonprofit educational trade association for the local area wireless technology industry. Its goal is to promote the wireless industry in general. To this end, WLANA acts as a clearinghouse for information about wireless local area applications, issues, and trends, wireless local area products, and wireless personal area products. You can find WLANA at `www.wlana.org/ index.html`.

✔ **Wi-Fi Alliance (formerly WECA):** Formed in 1999, the Wi-Fi Alliance is a nonprofit, international association that certifies the interoperability of wireless local area network products based on IEEE 802.11 specifications. By the spring of 2004, the Wi-Fi Alliance boasted over 200 member companies from around the world and had certified over 1,000 devices. All the equipment used in the making of this book was tested and certified by the Wi-Fi Alliance. You can find the Wi-Fi Alliance at `www.weca.net`.

Local Wireless Groups

More and more grassroots organizations are popping up across the world to offer free broadband Internet access. You can meet like-minded wireless fans and do some networking. The social kind of networking that is. You can find a user group in your area by checking out the following Web sites:

✔ **Practically Networked:** `www.practicallynetworked.com/tools/ wireless_articles_community.htm`

✔ **WirelessAnarchy:** `www.wirelessanarchy.com/#Community%20Groups`

✔ **The Certified Wireless Network Professional site's users group locator:** `www.cwne.com/wug/search.php`

✔ **WirelessCommunities:** `http://wiki.personaltelco.net/index. cgi/WirelessCommunities`

Other Industry Associations

We have included some other industry associations here because they have input (pun intended) into your wireless networks as well. Primarily, these groups make the client devices.

- **CompactFlash Association (CFA):** The CFA is a nonprofit, mutual-benefit corporation that maintains and promotes the CompactFlash and CF+ specification as a worldwide, small form factor, removable card standard. The CF+ specification adds additional card functionality, including Bluetooth, 802.11b wireless LAN, and wireless digital cell phone cards. You can find the CFA at `www.compactflash.org`.

- **Personal Computer Memory Card International Association (PCMCIA):** Founded in 1989, the PCMCIA is an international standards body and trade association with over 200 member companies. Its new mission includes standards for small form factor cards. The PC Card Standard contains all of the physical, electrical, and software specifications for the PC Card technology. You can find this organization at `www.pcmcia.org`.

- **Universal Serial Bus (USB) Implementers Forum, Inc.:** USB Implementers Forum, Inc., is a nonprofit corporation founded by the developers of the Universal Serial Bus specification as a support organization and forum for the advancement and adoption of USB technology. You can find them at `www.usb.org/home`.

Appendix B

Wireless Standards

. . . .

In This Chapter

▶ Understanding wireless standards

▶ Knowing which standard is right for your needs

. . . .

*T*here are so many standards now that it is difficult to keep up. A, B, G, sounds like an alphabet doesn't it? Well, in this Appendix, we explain the differing standards and provide you a reference for when you forget. And forget you will! Who knew there were so many of them?

Wireless networking standards are introduced through an organization called the IEEE, a nonprofit, technical professional association. The full name is the Institute of Electrical and Electronics Engineers, Inc.

802.1x

This standard allows for centralized authentication of wireless users or access points. It also allows multiple authentication algorithms and, as an open standard, lets vendors offer enhancements. 802.1x uses an existing authentication protocol known as the Extensible Authentication Protocol (EAP). EAP messages are encapsulated in 802.1x messages and referred to as EAPOL, or EAP over LAN, offering far greater security than WEP. EAP works by existing inside of PPP's (Point-to-Point Protocol) authentication protocol and provides a general framework that different authentication methods can use. It is supposed to prevent proprietary authentication systems and let everything from passwords to challenge-response tokens to public-key infrastructure certificates work within the same infrastructure.

802.11

802.11 is the granddaddy of wireless networking standards. The IEEE 802.11 specifications specify an "over-the-air" interface between a wireless client and a base station or access point, as well as at the peer-to-peer level among wireless clients. These standards can be compared to the IEEE 802.3 standard for Ethernet for wired LANs. They address two of the OSI layers, the Physical (PHY) layer and Media Access Control (MAC) sub-layer of the Data Link layer. The wireless standard is designed to resolve compatibility issues between manufacturers of Wireless LAN equipment.

This standard provides 1 or 2 Mbps transmission in the 2.4 GHz band using either frequency hopping spread spectrum (FHSS), direct sequence spread spectrum (DSSS), or infrared.

802.11a

After the finalization of the 802.11 standard, it became apparent that the 2 Mbps bit rate wouldn't cut it, especially compared to 10 Mbps Ethernet. Soon after, the IEEE started two taskgroups to work on this problem: 802.11a and 802.11b. The goal of the two groups was to define higher bit rate refinements to the 802.11 standard. Hence, 802.11a is an extension to 802.11 that provides up to 54 Mbps speed in the 5GHz band.

At the Physical layer, 802.11a uses an orthogonal frequency division multiplexing encoding scheme rather than FHSS or DSSS. It offers less potential for radio frequency (RF) interference because it operates in the 5GHz band. This standard is quickly gaining ground on 802.11b.

Even though approval for the 802.11a and 802.11b standards came at roughly the same time, it took a year longer for 802.11a equipment to hit the market due to the complexity of the standard. In fact, it took until late 2002 before 802.11a equipment hit critical mass.

802.11b

This extension provides 11 Mbps transmission (with a fallback to 5.5, 2, and 1 Mbps) in the 2.4 GHz band. You may hear people refer to 802.11b as Wi-Fi or Wireless-Fidelity: a throwback to Hi-Fi. One of the original wireless specifications in use, 802.11b uses only DSSS. It was first implemented in a 1999 ratification to the original 802.11 standard. 802.11b equipment is backward compatible to 802.11 equipment using DSSS. It was at the time considered a stopgap until the adoption of the 802.11a standard, but it soon became the dominant standard with the largest installed base. Most wireless solutions today either use or support this standard. It is also probably the least expensive to implement, although 802.11a is quickly catching up.

802.11c

You don't see or hear about this one much. 802.11c provides required information to ensure proper bridge operations. Product developers use this standard when developing access points, so most users will not notice it.

802.11d

802.11d is a little-known standard. The intent of 802.11d was to harmonize frequency and bandwidth around the world so that wireless equipment can interoperate. Enough said.

802.11e

802.11e is being defined to provide support for Quality of Service (QoS) traffic and thereby improve support of audio and video (such as MPEG-2) applications. Because 802.11e falls within the MAC layer, it will be common to all 802.11 PHYs and be backward-compatible with existing 802.11 wireless LANs.

802.11f

You don't see or hear much about 802.11f. 802.11f provides the necessary information that access points need to exchange in order to support the distribution system functions, like roaming. Without this standard, the IEEE recommends using similar vendors to support interoperability.

802.11g

The 802.11g standard defines the way wireless communicates at higher bit rates of up to 54 megabits per second while remaining backward-compatible with the 11 Mbps 802.11b standard. This bit rate enables streaming media, video downloads, and more users. 802.11g is gaining ground on the earlier 802.11a/b standards in industry and quickly becoming the more prevalent standard for wireless access.

802.11h

802.11h is a new specification that addresses the requirements of the European regulatory bodies. It provides for dynamic channel selection (DCS) as well as transmit power control (TPC) for devices operating in the 5GHz band, like the 802.11a specification does. In Europe, there is a greater need to avoid interference with satellite communications, which have "primary use" designations and can be interfered with by the 5 GHz band. This standard helps eliminate any potential for that interference.

802.11i

802.11i is the security panacea for wireless LANs. 802.11i incorporates stronger encryption techniques, such as AES (Advanced Encryption Standard). It is designed to improve on the weaknesses found in the existing WEP standards used by the other wireless standards. 802.11i includes strong encryption and a robust key management scheme. On the flipside, 802.11i will require new hardware chipsets, so it will not be compatible with existing hardware.

802.11j

The 802.11j task group has the mandate to refine some physical and data link issues for 5 GHz wireless networking with the view to the coexistence and eventual convergence of the IEEE 802.11a and European/Japanese HIPERLAN/2 standards.

802.11k

802.11k is another Quality of Service standard, but this one is for the Radio layer (physical). The mandate is to ensure the quality of service over an 802.11 link.

802.11n

802.11n is an effort to provide user throughput speeds of 100M bits/sec or more, with vendors like Agere pushing for 500M bits/sec. Current speeds in 802.11g for example, have data rates of 54M bits/sec which usually results in user throughput of considerably less, arguably around only 18 to 22M bits/sec. Originally anticipated January of 2004 it is still waiting approval.

802.15

In March 1998, the IEEE formed the WPAN Study Group. The study group's goal was to investigate the need for a wireless network standard for devices within a personal operating space (POS). In May of 1998, the Bluetooth Special Interest Group (SIG) formed. In March of 1999, the WPAN study group became IEEE 802.15, the WPAN Working Group. The 802.15 WPAN (Wireless Personal Area Network) is an effort to develop standards for Personal Area Networks or short distance wireless networks. These WPANs address wireless networking of portable and mobile computing devices, such as PCs, Personal Digital Assistants (PDAs), peripherals, cell phones, and pagers, letting these devices easily communicate with one another.

Since the formation of 802.15, three projects have started. The first (TG1) was the Bluetooth project that released the Bluetooth 1.0 Specification in July of 1999. The project will produce an approved IEEE standard derived from the Bluetooth standard. The second, or TG2, will address the issue of co-existence of 802.11 and 802.15 networks. Currently, Bluetooth networks create havoc with 802.11 networks. And the third, or TG3, will work on delivering a standard for high bit rate (20 Mbps or higher) WPANs.

802.16

The 802.16 standard is a broadband wireless standard for Wireless Metropolitan Area Networks (WirelessMAN or WMAN). This standard, also known as Broadband Wireless Access (BBWA), addresses the "first-mile/last-mile" connection in wireless metropolitan area networks, focusing on the efficient use of bandwidth between 10 and 66 GHz. Unless you are a large business, it is unlikely you'll deal with this standard.

Appendix C

The Fundamentals of Radio Frequency

*W*e cannot do justice to the discussion of radio frequency in one Appendix. It is the stuff of many books. Having said that, you must understand some concepts in order to set up and administer your WLAN. This Appendix provides a glimpse into the fascinating world of radio frequencies. You may want to peruse this appendix before calculating your link budget in Chapter 2.

Radio Frequency

When teaching networking, we often use the example of throwing a rock into a river to teach the concept of attenuation. Think of going down to the water and throwing in a rock. You see an epicenter where the rock went in and waves rippling out from that epicenter. The farther you get from the center, the weaker the waves get. The concentric circles that you see are similar to the radio waves as they propagate away from the antenna.

Radio frequencies are high-frequency (and in our, case ultra- and super-high frequency, as shown in Table C-1) alternating current (AC) signals passed along copper wire or some other conductor until an antenna radiates them into the air. The antenna transforms the wireless signal into a wired signal and vice versa. When the antenna propagates the high-frequency AC signal into the air, it forms radio waves. These radio waves propagate, or move away, from the source in a straight line in all directions. Just imagine the rock going into the water.

Table C-1	Radio Frequency Spectrum
Frequency	*Description*
Up to 300 Hz	Extremely Low Frequency (ELF)
300 Hz–3 kHz	Voice frequency
3 kHz–30 kHz	Very Low Frequency (VLF)
30 kHz–300 kHz	Low Frequency (LF)
300 kHz–3 MHz	Medium Frequency (MF)
3 MHz–30 MHz	High Frequency (HF)
30 MHz–300 MHz	Very High Frequency (VHF)
300 MHz–3 GHz	Ultra-High Frequency (UHF)
3 GHz–30 GHz	Super High Frequency (SHF)
30 GHz–300 GHz	Extremely High Frequency (EHF)

In the table, *Hz* denotes hertz. We use the term *hertz* to represent the unit for frequency. One hertz simply means one cycle (event) per second; 10 Hz means ten cycles (events) per second; and so on. You can apply hertz to any periodic event. For example, the clock speed of your Pentium might be said to tick at 2.2 GHz. The reciprocal of frequency is time (period): a frequency of 1 Hz is equivalent to a period of 1 second, and a frequency of 1 MHz is equal to a period of 1 microsecond. You should know some multiples, as follows:

Term	*Symbol*	*Equivalence*
1 kilohertz	kHz	10^3 Hz or 1,000 Hz
1 megahertz	MHz	10^6 Hz or 1,000,000 Hz
1 gigahertz	GHz	10^9 Hz or 1,000,000,000 Hz

This may or may not make sense, but either way, some examples cannot hurt. For example, standard domestic AC electric power (220v [volt] or 110v voltage) is 50–60 Hz. If you play music on the side, middle C is 261.625 Hz. If you don't play music but listen to it, FM radio broadcasts are 88–108 MHz. The clock speed of the Intel 4004, the world's first commercial microprocessor, was 104 kHz. Today's Pentium 4 has a clock speed of around 3 GHz.

The Federal Communications Commission (FCC) in the United States as well as other government agencies around the world have made parts of the spectrum available for unlicensed radio networks as long as they meet local regulations. Currently, these bands are the Industrial, Scientific, and Medical (ISM) band (operating at 2.4 GHz) and the U-NII (Unlicensed National Information Infrastructure) band (operating at 5.8 GHz).

In Appendix B, you can read about the various wireless standards. The 802.11b standard supports data rates of up to 11 Mbps, whereas 802.11a and 802.11g support data rates of up to 54 Mbps. The difference in data rates is caused by one of two things: more bandwidth or better encoding. The 802.11g standard works in the same 2.4 GHz band used by 802.11b but uses OFDM rather than DSSS and uses 64QAM rather than DQPSK. The 802.11a standard has more bandwidth.

64QAM and DQPSK? That's got to be Greek! No, it's *Geek* for modulation technology. Because we have digital bits to transmit over the air, our radio transceiver must convert the digital data to an analog signal. Converting digital data to analog signals is *modulation.* You can modulate data by using the amplitude, the frequency, or the phase of the signal (all three components of the signal). The term *shift keying* is sometimes substituted for the term *modulation,* and we use that term interchangeably here. Tables C-2 and C-3 provide the modulation techniques for 802.11, 802.11a, 802.11b, and 802.11g.

Table C-2	Modulation Techniques: 802.11, 802.11b, and 802.11g		
Spreading Code	*Modulation Technology*	*Data Rate*	
2.4 GHz DSSS	Barker Code	DBPSK	1 Mbps
		DQPSK	2 Mbps
	CCK	DQPSK	5.5 Mbps
		DQPSK	11 Mbps
2.4 GHz FHSS	Barker Code	2GFSK	1 Mbps
		4GFSK	2 Mbps

CCK: Complimentary Code Keying
DBPSK: Differential Binary Phase Shifting Key
DQPSK: Differential Quadrature Phase Shifting Key
GFSK: Gaussian Phase Shifting Key

Table C-3	Modulation Techniques: 802.11a
Modulation Technology	*Data Rate*
BPSK	6 Mbps 9 Mbps
QPSK	12 Mbps 18 Mbps

(continued)

Table C-3 *(continued)*

Modulation Technology	Data Rate
16QAM	24 Mbps
	36 Mbps
64QAM	48 Mbps
	54 Mbps

BPSK: Binary Phase Shifting Key
QPSK: Quadrature Phase Shifting Key
QAM: Quadrature Amplitude Modulation

The 802.11g standard achieves its high data rates through the use of the Quadrature Amplitude Modulation (QAM) technique. With QAM, there are 12 phase angles with 2 different amplitudes. Eight phase angles have a single amplitude, and four have two amplitudes, resulting in 16 different combinations. QAM uses each signal change to represent 4 bits. Consequently, the data rate is four times the baud rate.

You may find vendors who support 802.11a, b, and g in a single device. We cover that in Chapter 3.

Early in this Appendix, we talk about the pebble creating a splash in the water and waves rippling out until they dissipate. We call this phenomenon *attenuation.* Figures C-1 and C-2 demonstrate attenuation for outdoors and indoors by using the 802.11b standard as an example. You can see that the signal travels farther outdoors because no walls, floors, or any other obstructions absorb, reflect, refract, or diffract the signal.

Figure C-1 shows what you would expect in theory. The access point does not radiate perfect circles at precisely these distances in the real world. In reality, the radiation patterns are more oblong and flatter than a circle. Popular wisdom holds that 802.11b technology, which is generally held to be effective up to about 300 feet, offers better coverage than 802.11a equipment. Theoretical calculations put 802.11a coverage at roughly one-fourth of that range. However, tests show that 802.11a operates with acceptable reliability to well over 200 feet. Moreover, throughout most of its range, including the maximum, it offers a throughput advantage over 802.11b at the same distances. Products based on 802.11a use the 5.8 GHz band.

Physics dictate that higher frequencies have a larger path loss (greater spatial attenuation) — and therefore, shorter range than lower frequencies — when all other variables are the same. Thus, the 802.11g products have a greater range than 5 GHz products for the same data rate. But the current higher susceptibility to interference in the 2.4 GHz band might affect the range of 802.11g products more in noisy and congested environments than products in the 5 GHz band.

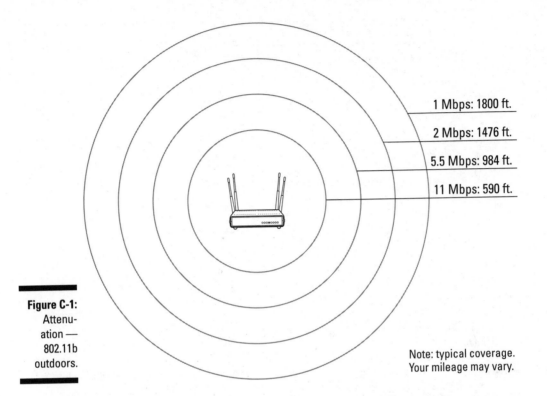

1 Mbps: 1800 ft.

2 Mbps: 1476 ft.

5.5 Mbps: 984 ft.

11 Mbps: 590 ft.

Figure C-1:
Attenu-
ation —
802.11b
outdoors.

Note: typical coverage.
Your mileage may vary.

Currently the 5 GHz band is cleaner from interference than the 2.4 GHz band. However, both bands are unlicensed. With the emergence of products creating interference in the 5 GHz band (at least three cordless telephone products workin the 5.8 GHz band, representing the top four 802.11a channels), interference may eventually affect 802.11a products much like 802.11b and g products. At the same time, the 5 GHz band has more bandwidth than the 2.4 GHz band for unlicensed devices, and thus there is more room to avoid such interference.

Specific implementation details of different vendors, such as power output, receiver sensitivity, antenna design, and other factors will also affect the range.

Another consideration is the number of usable channels. 802.11b (or Wi-Fi) is limited to three clear channels. When you deploy more than three contiguous cells, you likely will find some performance degradation (up to as much as 50 percent) because of co-channel interference (CCI) between cells operating on a given channel. With Wi-Fi, there's no way to avoid duplication of channel usage more than one cell diameter away. And the closer together the cells, the more interference. Table C-4 shows the various channels in use for 802.11b and g and their frequencies. Channels 1, 6, and 11 are the non-overlapping channels.

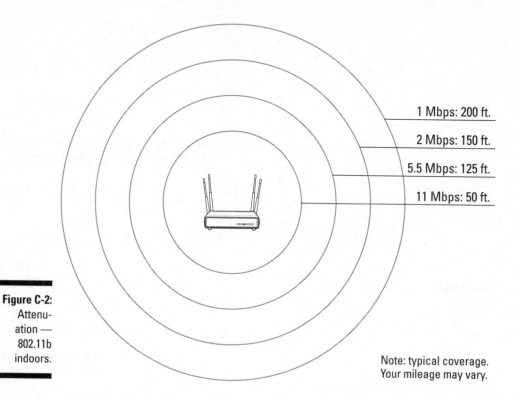

1 Mbps: 200 ft.

2 Mbps: 150 ft.

5.5 Mbps: 125 ft.

11 Mbps: 50 ft.

Figure C-2:
Attenu-
ation —
802.11b
indoors.

Note: typical coverage.
Your mileage may vary.

Table C-4	Frequency and Channels		
Channel	*Frequency*	*Channel*	*Frequency*
1	2.412 GHz	8	2.447 GHz
2	2.417 GHz	9	2.452 GHz
3	2.422 GHz	10	2.457 GHz
4	2.427 GHz	11[1]	2.462 GHz
5	2.432 GHz	12	2.467 GHz
6	2.437 GHz	13[2,3]	2.472 GHz
7	2.442 GHz	14[4]	2.477 GHz

[1] *North America uses channels 1–11*
[2] *Europe (except France) uses channels 1–13*
[3] *France uses channels 10–13*
[4] *Japan uses channels 1–14*

Table C-5 lists the 802.11a channels. With 802.11a's eight channels, however, you can easily arrange a cell grid so that access points using the same channel are at least twice as far apart — and the overall density of cells using any given channel is roughly one-fourth as great. This should greatly reduce the effect of CCI, if not eliminate it altogether.

Table C-5	802.11a Usable Channels	
Frequency Band	*Channel Number*	*Center Frequencies (GHz)*
U-NII Lower Band (5.15–5.25 GHz)	36	5.180
	40	5.200
	44	5.220
	48	5.240
U-NII Middle Band (5.25–5.35 GHz)	52	5.260
	56	5.280
	60	5.300
	64	5.320
U-NII Upper Band (5.725–5.825 GHz)	149	5.745
	153	5.765
	157	5.785
	161	5.805

An access point can support all standards because ultimately they are complementary. As previously mentioned, the 802.11b and 802.11g standards operate in the Industrial, Scientific, and Medical (ISM) band, and 802.11a operates in the Unlicensed National Information Infrastructure (U-NII) band. So, 802.11g can complement 802.11a by adding three additional channels in the 2.4 GHz band to existing 802.11a channels. This creates more network capacity to allow for additional users. Both technologies have advantages that when used in combination, offer an even stronger product. Another advantage of 802.11a is that the 5 GHz base has more capacity around the world. Currently, there are 13 channels in North America (including U-NII and ISM bands), 8–19 channels in Europe, and 5–12 channels in Asia. The more channels you have, the more aggregate throughput you can have.

However, although there are up to 14 allocated channels for the 802.11b and g standards, there are only 3 non-overlapping channels. (That is, only three APs could work in one area without interfering with each other.) In the 802.11a standard, all the channels are non-overlapping. (That is, channels don't interfere with each other at all.) 802.11a has 12 non-overlapping channels: 8 dedicated to indoor and 4 to point-to-point. Remember, this helps ensure less interference.

That's a lot of numbers and standards to remember. We summarize the transfer method, frequency, and data rates for the various standards for you in Table C-6.

Table C-6		Standards and Values	
Standard	*PHY Transfer Method*	*Frequency Band*	*Data Rates (Mbps)*
802.11 legacy	FHSS, DSSS, IR	2.4 GHz, IR	
802.11b	DSSS, HR-DSSS	2.4 GHz	
"802.11b+" non-standard	DSSS, HR-DSSS (PBCC)	2.4 GHz	1, 2, 5.5, 11, 22, 33, 44
802.11a	OFDM	5.2, 5.5 GHz	
802.11g	DSSS, HR-DSSS, OFDM	2.4 GHz	

Behavior of Radio Waves

Here are some simple concepts that are necessary to your understanding of RF and hence your ability to design, install, and administer your wireless network. You need to understand the concept of gain, loss, reflection, refraction, diffraction, scattering, absorption, and free space loss.

Gain

Gain describes an increase in a radio frequency signal's amplitude. Usually, gain is an active process. This means that you can use an external power source (such as an antenna) to amplify a signal, or you can use a high-gain antenna to focus the beam width of a signal to increase its amplitude.

But passive processes can also cause gain. For example, reflected signals (see the upcoming section, "Reflection") can combine with the main signal to increase the signal's strength. You must know how to measure gain in your system to find the signal strength at your client or to know whether you're violating any laws regarding signal strength or power levels.

Loss

Conversely, *loss* is a decrease in signal strength. Many factors can cause signal loss. For instance, resistance in connectors and friction in a cable can cause signal loss (*attenuation*), hence the reason for maximum cable runs.

Mismatching impedance in cables and connectors can cause loss by reflecting signal back toward the source. Objects directly in the path of the radio waves can absorb, reflect, or disrupt RF signals, thus causing degradation. Also, free space loss results from sending a RF signal over the air: The further you go, the weaker the signal gets. Like with gains, you must know how to quantify losses. You need to know when you are approaching the sensitivity threshold for your receiver. The sensitivity threshold is the point where the receiver can clearly distinguish the signal from background noise. To find out how to calculate gains and losses, keep reading.

Reflection

Reflection is the phenomenon of a propagating wave being thrown back from a surface. Reflections result from the surface of the Earth, buildings, walls, and other large objects. When the surface is smooth, the reflected signal may remain intact, but some absorption and scattering of the signal is likely. The reflection of the signal from many objects at once causes multipath, which we deal with in Chapter 13. *Multipath* can cause signal degradation or cancellation.

Refraction

Refraction occurs when sound wave changes mediums. (The wave bends as it passes through a medium of different density.) As a RF wave passes into a denser medium (such as a trough of cold air lying in a valley), the wave bends, so its direction changes. Some of the wave reflects away from the intended signal path, and some bends in another direction altogether. Refraction is mostly a concern when dealing with longer links. Atmospheric conditions might bend the signal away from the intended receiver.

Diffraction

Diffraction is the apparent bending of light waves around obstacles in its path. It occurs when an object with a rough or irregular surface obstructs the radio path. Diffraction is the effect of waves bending or turning around the obstruction. We can explain diffraction by looking at our pebble in the pond example again. Suppose when you threw a pebble into water, a small stick was stuck in the muck. (Say that real fast five times.) The waves rippling out hit the stick. The stick blocks some waves, but most bend around the stick.

Scattering

Scattering occurs when the wave passes through a medium consisting of many small objects compared with the wave itself. Rough surfaces, small objects, or other irregularities such as foliage, street signs, or streetlights can cause the signal to scatter. Scattering can destroy the main signal.

Absorption

Absorption occurs when a RF signal strikes an object and does not pass through, reflect, or diffract around the object. The object absorbs the incoming signal. The curtains on your windows or the padded cubicle dividers can absorb your signals.

Free space loss

As the wave propagates away from the source, it loses steam and eventually peters out. This phenomenon is *free space loss* and is similar to attenuation in a copper cable. Later in this Appendix, you can see how to calculate how much you lose as you move the receiver farther and farther away from the source.

Fresnel zone

Before looking at RF units of measure, you need to understand one more concept about waves: the Fresnel zone. The *Fresnel zone* occupies a series of concentric, ellipsoidal areas around the line-of-sight (LOS) path. This area is important because it defines an area about the LOS that you should ensure is not blocked. Trees, towers, buildings, and other solid objects in the Fresnel zone can absorb, scatter, reflect, or diffract a signal and cause degradation. Typically, 20–40 percent blockage in the Fresnel zone introduces little or no interference of the signal. Err on the conservative side, and aim for no more than 20-percent blockage of the zone.

If you think free space loss and Fresnel zones are fun, you might get your kicks from the paper, "VHF/UHF/Microwave Radio Propagation: A Primer for Digital Experimenters," at www.tapr.org/tapr/html/ve3jf.dcc97/ve3jf.dcc97.html.

RF Units of Measure

When you build a simple home network, chances are that you'll just buy the equipment, plug it in, configure it, and use it. You're like a kid on his birthday: You can't wait to play with your new toys. On the other hand, when you choose to implement a network for an SMB or an enterprise where you need to fool with external antennas, cables, bridges, and repeaters, you need some basic knowledge about power and signal levels. This section serves as your introduction.

Watt's that, you say?

The basic unit of power is the *watt*. The watt (W) is defined as one ampere (A) of current at one volt (v). In other words, one watt is equal to one ampere multiplied by one volt. Thinking of a garden hose can help you understand these concepts: The water flow is the amperes (or current), and the pressure in the hose is the voltage (or electrical circuit). The FCC allows only 4 watts of radiated power from an antenna in a point-to-multipoint wireless LAN connection using unlicensed 2.4 GHz spread spectrum equipment. This is likely your configuration: one access point and many clients.

Now, 4 watts might not seem like a lot, but consider the typical night light. (You know the one, the light that helps you find your way to the john in the night.) It is about 7 watts. On a clear night with no light pollution, you can spot a 7-watt light from 50 miles (or approximately 83 kilometers) away in all directions. Now imagine that you could encode this light to send data. You would have a wireless network. So, 4 watts allows you to create an effective and efficient wireless local area network.

Generally, when working with WLANs and WPANs, you can use power levels as low as 1 milliwatt (mW, 1/1000 of a watt). Power levels on a WLAN segment are rarely above 100 mW, which is sufficient to transmit up to half of a mile (0.83 km) under ideal conditions. Access points generally radiate between 30 and 100 mW of power, depending on the particular manufacturer. When you look at wireless equipment, you'll see power levels specified as either mW or dBm (covered next). Both measurements represent an absolute amount of power.

I hear 'bels

Sensitive receivers can pick up signals as small as 0.000000001 watts. This is a very small number that most likely means very little to you. Instead of using very small absolute numbers, you can use another measure that makes the numbers more meaningful. This measure is the *decibel*, based on a logarithmic relationship to the watt. The decibel (dB) is a measure of relative power or signal strength. However, it is not an absolute measure, like a watt for power or a volt for signal.

The reference point between the logarithmic dB scale and the linear watt is

```
1 mW = 0 dBm
```

The lowercase *m* in *dBm* refers to the reference point of 1 milliwatt (mW) rather than one watt. dBm is a measure of absolute power and ir not relative. It measures the power relative to one milliwatt. You calculate dBm as follows:

```
dBm = 10 log(mW/1)
```

or

```
dBm = 10 log(W/.001)
```

where mW is the signal in milliwatts.

Table C-7 provides a conversion chart for you, so you don't have to do the math. The first entry shows that 0 dBm is equal to 1 mW. You can calculate these values by popping the formula into Google. Google likes logs, and not just the kind that tells them what you did on that site. For those of you who don't want to use Google to try logs, you can use www.bessernet.com/jobAids/dBmCalc/dBmCalc.html to do the conversion for you.

Table C-7		dBm to Power Conversion Chart	
dBm	*Power*	*dBm*	*Power*
0	1 mW	20	100 mW
1	1.25 mW	21	125 mW
2	1.6 mW	22	160 mW
3	2 mW	23	200 mW
4	2.5 mW	24	250 mW
5	3.2 mW	25	320 mW
6	4 mW	26	400 mW
7	5 mW	27	500 mW
8	6.4 mW	28	640 mW
9	8 mW	29	800 mW
10	10 mW	30	1 W
11	12.5 mW	31	1.25 W
12	16 mW	32	1.6 W
13	20 mW	33	2 W
14	25 mW	34	2.5 W
15	32 mW	35	3.2 W
16	40 mW	36	4 W
17	50 mW	37	5 W
18	64 mW	38	6.4 W
19	80 mW	39	8 W

We measure power gains and losses in decibels, not in watts. This should make sense to you because gains and losses are relative concepts, and decibels are relative measures.

For measuring gains and losses, you use dB except when measuring the gain of an antenna, when you use the term dBi. The *i* in *dBi* stands for dB gain over an isotropic antenna. This means that you compare the change in power with the mythical isotropic antenna. An *isotropic antenna* is an ideal one that sprays useful radio waves in all directions, including up and down, with equal intensity, at 100-percent efficiency. Our Sun is an isotropic radiator. (Tell that to someone who lives north of the 49th parallel.) Therefore, dBi is a decibel gain realized by a gain antenna, compared with what the theoretical isotropic antenna would do with the same power level. Thus, dBi is a relative measurement.

RF Mathematics

In Chapter 2, we show you how to do a site survey. To complete the site survey, you need to do a link budget. This section provides all the formulas that you need to do your link budget. If you want to really dive into radio waves, you should understand the theory behind the formulas. For the rest of us, just know they are here and that you need only substitute the correct values.

Calculating decibels

So, when a signal loses 3 dB, is that a lot? A 3 dB loss indicates that the signal lost half of its power. As they say in math, QED (*quod erat demonstrandum: that is, that which was to be proved*):

$$dB = 10 \log_{10}(P2 / P1)$$

$$-3 \ dB = 10 \log_{10}(P2 / 100)$$

$$-0.3 = \log_{10}(P2 / 100)$$

$$10 - 0.3 = P2 / 100$$

$$0.50 = P2 / 100$$

$$P2 = 50\%$$

Now, you may not feel like using a slide rule (or Google) to calculate dB, but you can use the handy following rules. We have the 10s and 3s of RF math as follows:

1. –3 dB = ½ the power in mW

2. +3 dB = 2 times the power in mW

3. –10 dB = ⅒ the power in mW

4. +10 dB = 10 times the power in mW

When calculating gains or losses, factor the numbers by 3 or 10 or both to get a quick value for loss or gain.

Decibel losses (and gains) are additive. If you have an access point connected to a cable with a –2 dB loss and a connector with a –1 dB loss, you have a –3 dB loss, or half the power radiated by the access point. You need to remember this additive rule when calculating link budgets.

Suppose you have a 33 dBm gain (+33 dBm). You can break 33 down into 10 + 10 + 10 + 3. Remember the handy rules that state that +10 is equivalent to 10 times and that +3 is equivalent to 2 times. You know that 1 mW is equal to 0 dBm, so this is where you start. You can calculate that +33 dBm equals 2 watts, as follows:

$1 \text{ mW} \times 10 = 10 \text{ mW}$

$10 \text{ mW} \times 10 = 100 \text{ mW}$

$100 \text{ mW} \times 10 = 1000 \text{ mW}$

$1000 \text{ mW} \times 2 = 2000 \text{ mW}$ or 2 watts

Consider a negative example. Suppose that you have –23 dBm. You can break 23 down into –10 + –10 + –3. You know that 1 mW is equal to 0 dBm, so this is where you start. You can calculate that –23 dBm equals 5 microwatts as follows:

$1 \text{ mW}/10 = 100 \text{ }\mu\text{W}$

$100 \text{ }\mu\text{W}/10 = 10 \text{ }\mu\text{W}$

$10 \text{ }\mu\text{W}/2 = 5 \text{ }\mu\text{W}$

We both bought +16 dBi antennas from Hugh Pepper (http://mywebpages.comcast.net/hughpep). If you apply 1 watt of power, the output power at the antenna is as follows:

$1 \text{ W} + 16 \text{ dBi} = 40 \text{ W}$

This calculation works exactly like dB calculations. This means that a 16 dBi gain is +10 + 3 + 3 or 10 times, and 2 times, and 2 more times. Logically, antennae do not degrade the signal (assuming they're working right), so dBi is always a gain.

Calculating path loss

You can calculate the path loss by using one of the following formulas:

```
Lp = 32.4 + 20 logf + 20 logd (kilometers)
```

```
Lp = 36.6 + 20 logf + 20 logd (miles)
```

where L_p is the path loss given for either kilometers or miles, *f* is the frequency in MHz, and *d* is the distance in kilometers in the first formula and miles in the second. The path loss for a one-mile path is figured as follows:

$$L_p = 36.6 + 20 \log 2437 + 20 \log 1 \text{ (kilometers)}$$
$$L_p = 36.6 + (20 \times 3.39) + (20 \times 0) = 104.4 \text{ (miles)}$$

Using Channel 6 across a one-mile path, the loss is 104.4 dB. That's quite a lot!

That formula is pretty intimidating, so we provide Table C-8 as an estimate of path loss for 2.4 GHz networks.

Table C-8	Free Space Loss	
Distance (In Meters)	*Distance (In Feet)*	*Loss (In dB)*
100	328.08	80.23
200	656.17	86.25
500	1,640.42	94.21
1,000	3,280.84	100.23
2,000	6,561.68	106.25
5,000	16,404.20	114.21
10,000	32,808.40	120.23

Calculating antenna length

You can calculate the length of the antenna by using the following formula:

```
L = 984 / f
```

where f is the frequency in MHz, and the result L (length) is in feet.

$L = 984 / 2412$

$L = 0.4079$ feet or 4.895 inches

For the 2.4 GHz ISM band, the ideal antenna length is 4.89 inches.

Calculating coaxial cable losses

Coaxial cable eats signal strength for breakfast. One very common coax cable (the Times Microwave LMR 240) has a 12.7 dB loss for 100 feet. Think of that — a –3 dB loss means the that power is halved! A –6 dB loss means that the power is halved again, or one-fourth of the power. At –9 dB, you have but one-eighth left; at –12 dB, you have only one-sixteenth of the power remaining.

Cable losses increase linearly. Or in other words, when you lose 12 dB per 100 feet, you lose 6 dB per 50 feet, and so on. Using the LMR 240 coaxial cable as an example, to calculate the loss over 10 feet, the calculation is

$10/100 \times 12.7 = 1.27$ dB

Over 15 feet, it is

$15/100 \times 12.7 = 1.91$ dB

You can calculate the power ratio associated with the dB value by dividing the dB value by 10 and raising $\frac{1}{10}$ to that power. For example

$1/10^{0.127} = 0.746$

This means that you'll have only 75 percent of your input power at the end of the 10-foot run of cable. It goes without saying that you should keep your cable runs to the bare minimum.

Calculating the Fresnel zone

You can calculate the radius of the Fresnel zone at the widest point by using the formula:

$$R = 43.3 \times \sqrt{d} / 4f$$

where d is the distance of the link in miles, f is the frequency in GHz, and the result R is in feet.

$R = 43.3 \times \sqrt{2}/4 * 2.4$

$R = 19.76$ feet

In the preceding solution, you have an 802.11b or g link that is 2 miles long. So take the visual line of site and measure about a 20 foot radius around the line: That is your Fresnel zone. Keep it clear!

Throw all this together. Figure C-3 shows an RF circuit with an access point, connectors, cables, and an antenna. The access point radiates 100 mW of power. Each connector adds 3 dB of loss. Each cable run loses 3 dB as well. Finally, the antenna is a 16 dBi gain antenna. Table C-8 shows the results of this circuit.

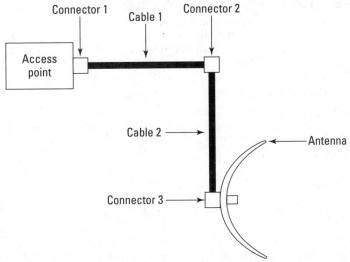

Figure C-3:
RF Circuit.

Table C-8	RF Circuit Power Calculation		
	Power Value	*Factor*	*Result*
Access point	100 mW		
Connectors	–9 dB	÷ 2 ÷ 2 ÷ 2	12.5 mW
Cables	–6 dB	÷ 2 ÷ 2	3.125 mW
Antenna	+16 dBi	× 10 × 2 × 2	125 mW

With an access point of 100 mW

Connectors: –9 dB / 2 / 2 / 2 = 12.5 mW

Cables: –6 dB / 2 / 2 = 3.125 mW

Antenna: +16 dBi × 10 × 2 × 2 = 125 mW

The result is 125 mW radiated power at the antenna. If these calculations make you dizzy, do not despair. The nice Community Wireless Project: Norfolk people have generously provided online calculators at `www.retro-city.co.uk/bovistech/wireless/calcs.htm`. They give you calculators for milliwatts to dBm (and vice versa), transmit power, operating system margin, and Fresnel zone.

Calculating the measurements for a home-grown antenna

Gotcha. We could give you all the mathematics to build your own antenna, but we won't. We prefer to point you toward the excellent sites that already do this. Try the following:

- `www.oreillynet.com/cs/weblog/view/wlg/448`
- `www.freeantennas.com`
- `www.turnpoint.net/wireless/cantennahowto.html`
- `www.netscum.com/~clapp/wireless.html`
- `www.ashtec.dyndns.org/ashtec/mods/index.html`

If you want a more comprehensive list, try `www.wirelessanarchy.com/#Antenna`.

That's it. You now have a basic understanding of the fundamentals of radio frequency. If you are interested in getting more information, check out the following sites:

- `www.sss-mag.com/ss.html`
- `http://hyperphysics.phy-astr.gsu.edu/hbase/ems1.html`
- `http://imagine.gsfc.nasa.gov/docs/science/know_l1/emspectrum.html`
- `www.alvarion.com/RunTime/Materials/PDFFiles/FHvsDS-ver7.pdf`
- `www.glenbrook.k12.il.us/gbssci/phys/Class/waves/u10l3b.html`
- `http://csep10.phys.utk.edu/astr162/lect/light/spectrum.html`
- `www.5ivenetworks.com/index2.asp?act=tool`

Who knows, may be you'll go on to become a Certified Wireless Network Administrator (`www.cwne.com`).

Index